Ask, Explore, Write!

Discover how to effectively incorporate literacy instruction into your middle or high school science classroom with this practical book. You'll find creative, inquiry-based tools to show you what it means to teach science with and through writing, and strategies to help your students become young scientists who can use reading and writing to better understand their world.

Troy Hicks, Jeremy Hyler, and Wiline Pangle share helpful examples of lessons and samples of students' work, as well as innovative strategies you can use to improve students' abilities to read and write various types of scientific nonfiction, including argument essays, informational pieces, infographics, and more. As all three authors come to the work of science and literacy from different perspectives and backgrounds, the book offers unique and wide-ranging experiences that will inspire you and offer you insights into many aspects of the classroom, including when, why, and how reading and writing can work in the science lesson.

Featured topics include:

- Debates and the current conversation around science writing in the classroom and society.
- How to integrate science notebooks into teaching.
- Improving nonfiction writing by expanding disciplinary vocabulary and crafting scientific arguments.
- Incorporating visual explanations and infographics.
- Encouraging collaboration through whiteboard modeling.
- Professional development in science and writing.

The strategies are all aligned to the Next Generation Science Standards and Common Core State Standards for ease of implementation.

From science teachers to curriculum directors and instructional supervisors, this book is essential for anyone wanting to improve interdisciplinary literacy in their school.

Troy Hicks is a professor of English and Education at Central Michigan University, and Director of the Chippewa River Writing Project. He has authored and co-authored nine books and over 30 journal articles and book chapters for teachers and other educators.

Jeremy Hyler is a middle school English teacher and a teacher consultant for the Chippewa River Writing Project, a satellite site of the National Writing Project. He is also a regular blogger for MiddleWeb.

Wiline Pangle is a lecturer in the Department of Biology at Central Michigan University. She also works with science teachers across Michigan to develop inquiry-based activities to promote sciences at all levels of education.

Other Eye On Education Books Available from Routledge
(www.routledge.com/eyeoneducation)

From Texting to Teaching
Grammar Instruction in a Digital Age
Jeremy Hyler and Troy Hicks

Create, Compose, Connect!
Reading, Writing, and Learning with Digital Tools
Jeremy Hyler and Troy Hicks

Matching Reading Data to Interventions
A Simple Tool for Elementary Educators
Jill Dunlap Brown and Jana Schmidt

Content Area Literacy Strategies that Work
Do This, Not That!
Lori G. Wilfong

Passionate Learners, 2nd Edition
How to Engage and Empower Your Students
Pernille Ripp

Passionate Readers
The Art of Reaching and Engaging Every Child
Pernille Ripp

Rigor in the K – 5 ELA and Social Studies Classroom
A Teacher Toolkit
Barbara R. Blackburn and Melissa Miles

The Common Core Grammar Toolkit
Using Mentor Texts to Teach the Language Standards in Grades 9–12
Sean Ruday

The First-Year English Teacher's Guidebook
Strategies for Success
Sean Ruday

Culturally Relevant Teaching in the English Language Arts Classroom
Sean Ruday

Essential Truths for Teachers
Danny Steele and Todd Whitaker

Ask, Explore, Write!

An Inquiry-Driven Approach to Science and Literacy Learning

Troy Hicks
Jeremy Hyler
Wiline Pangle

Taylor & Francis Group

NEW YORK AND LONDON

First published 2020
by Routledge
52 Vanderbilt Avenue, New York, NY 10017

and by Routledge
2 Park Square, Milton Park, Abingdon, Oxon, OX14 4RN

Routledge is an imprint of the Taylor & Francis Group, an informa business

© 2020 Taylor & Francis

The right of Troy Hicks, Jeremy Hyler and Wiline Pangle to be identified as authors of this work has been asserted by them in accordance with sections 77 and 78 of the Copyright, Designs and Patents Act 1988.

All rights reserved. No part of this book may be reprinted or reproduced or utilised in any form or by any electronic, mechanical, or other means, now known or hereafter invented, including photocopying and recording, or in any information storage or retrieval system, without permission in writing from the publishers.

Trademark notice: Product or corporate names may be trademarks or registered trademarks, and are used only for identification and explanation without intent to infringe.

Library of Congress Cataloging-in-Publication Data
Names: Hicks, Troy, author. | Hyler, Jeremy, author. | Pangle, Wiline, author.
Title: Ask, explore, write! : an inquiry-driven approach to science and literacy learning / Troy Hicks, Jeremy Hyler, Wiline Pangle.
Description: New York, NY : Routledge, 2020. | Includes bibliographical references. |
Identifiers: LCCN 2019044310 (print) | LCCN 2019044311 (ebook) | ISBN 9780367225124 (hardback) | ISBN 9780367225131 (paperback) | ISBN 9780429275265 (ebook)
Subjects: LCSH: Scientific literature—Study and teaching (Secondary) | Technical writing—Study and teaching (Secondary) | Science—Study and teaching (Secondary) | Language arts (Secondary) | Inquiry-based learning.
Classification: LCC Q225.5 .H53 2020 (print) | LCC Q225.5 (ebook) | DDC 507.1/2—dc23
LC record available at https://lccn.loc.gov/2019044310
LC ebook record available at https://lccn.loc.gov/2019044311

ISBN: 978-0-367-22512-4 (hbk)
ISBN: 978-0-367-22513-1 (pbk)
ISBN: 978-0-429-27526-5 (ebk)

Typeset in Palatino
by Swales & Willis, Exeter, Devon, UK

Troy's Dedication

I dedicate my work to my dad, Ron Hicks, who encouraged me to ask, explore, and write throughout my own education, and continues to push me to do so today.

Jeremy's Dedication

I dedicate my work to HB who taught me to keep fighting regardless of the situation. I also dedicate my work to Jeremy Winsor who makes me better every single day as a teacher. Thank you both.

Wiline's Dedication

I dedicate my work to my family, both at home and abroad.

Contents

Preface .... *xiii*
    A Note About the Companion Website .... *xiv*
    Link for the Companion Website.. *xv*
Meet the Authors .... *xvii*

Introduction: Our Science/Literacy Stories xix
    Jeremy's Story: My Path to Becoming a Mad Scientist...err,
    Science Teacher xxi
    Wiline's Story: From France to Michigan, with Many Stops in
    the Field xxiv
    Troy's Story: Stepping Into Science Literacy xxvi
    Next Steps xxviii

Chapter 1: Writing in Science?(!)..1
    Jeremy's Journey: My Introduction to (Not) Writing across the
    Curriculum 2
    A Bit of Scientific Writing History 6
        Wiline's Perspective: The Dreaded Science Textbook 7
        Troy's Perspective: A Deeper Look at WAC, WID, WTL, and
        WID Movements 9
        A Move Toward Disciplinary Literacy 11
        Strategies for Writing in Science 12
        Literacy Now in the NGSS 15
    Four Key Shifts Driving the Science/Writing Conversation 17
        1. Literacy and Science *Do* Mix 18
        2. Writing and Science are Both Process-Oriented 19
        3. Inquiry Matters 19
        4. Technology, Writing, and Science are Intertwined 20
    Conclusion: More Alike than Different 21

Chapter 2: Science Notebooks 23
    What Science Notebooks Are 24
    Why Use Science Notebooks 26
        Jeremy's Perspective 26
        Wiline's Perspective 28
    How to Teach with Science Notebooks 30
        Jeremy's Perspective: A Case Study with Photosynthesis 30
        Wiline's Perspective 35

       Observational Entries 35
       Data Recording Entries 38
       Brainstorming Entries 41
       Notetaking Entries 41
  How to Assess Science Notebooks 44
  Next Steps with Science Notebooks 48

Chapter 3: Visual Explanations with Infographics 51
  What Infographics Are 53
  Why Use Infographics 55
  How to Teach with Infographics 56
      Jeremy's Perspective: Introducing Infographics to Students 57
      Wiline's Perspective: A Deeper Dive into Creating Tables and
        Other Graphs 62
          Step 1: Organizing Data into Tables 63
          Step 2: Organizing Data into Graphs 64
              Step 2.1: Organizing Data into Line Graphs 65
              Step 2.2: Organizing Data into Pie Graphs 68
              Step 2.3: Organizing Data into Bar Graphs 68
          Considering the Benefits and Constraints of Various
             Graph Types 70
  How to Assess Infographics 71
  Next Steps with Infographics 73

Chapter 4: Encouraging Collaboration through Whiteboard Modeling 75
  Welcome to a Classroom with Students Engaged in Modeling
    Pedagogy 75
  How to Teach with Modeling 79
      What Happens with Whiteboarding: Preparing for, Conducting,
        and Debriefing a Board Meeting 79
          Before the Board Meeting: Working in Small Groups 80
          During the Board Meeting: Listening Attentively 80
          After the Board Meeting: Reflecting on Our Learning 81
  Jeremy's Perspective: Modeling in Middle School 81
      Wiline's Perspective: Whiteboarding with Undergraduates 85
  How to Assess Modeling and Whiteboarding 88
  Wiline's Perspective: 89
  Next Steps with Modeling and Whiteboarding 91

Chapter 5: Additional Strategies to Encourage Inquiry, Reading,
and Writing 95
  Moving into Deeper Inquiry with the Question Formulation
      Technique 96

Creating (Creative) Nonfiction 99
    Creative Nonfiction Example 1: Narrative Poetry 101
    Creative Nonfiction Example 2: Real Estate Ads 101
Jeremy's Perspective: Everyday Vocabulary and Science
    Vocabulary 104
Wiline's Perspective: Concept Mapping 105
Wiline's Perspective: Writing Strategies for Scientific Argument 108
    Pre-writing: Outlining 108
    Writing: Drafting 112
    Writing: Editing 112
    Peer Review 113
Literature that Uses Science 115
Apps, Websites, and Devices Worth Considering in the Science
    Classroom 116
Conclusion: Foreground Literacy Practices During Inquiry Activities 119

Chapter 6: Professional Development in Science and Writing..................121
The Design of the Beaver Island Institute for Science and Literacy
    Teachers 122
The Field Experiences 125
    Field Experience 1: Snake Boards (Observational Design) 125
    Field Experience 2: Line Transects/Botany 127
    Field Experience 3: Stream sampling 129
    Field Experience 4: Trail Cameras 130
    Field Experience 5: Removal of Plant Invasive Species 132
    Field Experience 6: Animal Behavior 135
Capturing and Interpreting Data: A Closer Look 137
Outcomes and Implications 139
    Participant Responses via Institute Surveys 139
    Sample Units 141
        Clean Water: Creation and Access 141
        Interactions and Interdependence 142
        Species Integration and Management 144
Rethinking Professional Learning for Scientific Literacy 144
Conclusion: Next Steps with the Next Gen Science Standards and
    Literacy Learning..146
Potential Next Moves for Teachers..147

References *149*

Preface

Through our work over four summers in an intensive, one-week professional development experience that integrates science and literacy learning, the three of us have had many opportunities to think about, plan out, and reflect on what it means to teach science with and through writing. From our work together hosting this event, the annual Beaver Island Institute, which we will describe in more detail later on, the ideas for this book arose.

Our collaboration on this book extends upon work that Jeremy and Troy had already begun in their previous books (*Create, Compose, Connect!* and *From Texting to Teaching*) and welcomes Wiline's voice into the conversation. As our reader, who more than likely is an educator who teaches science and writing in some manner, we welcome you into this dialogue, and encourage you to share your ideas with us through social media and by visiting our companion website.

Our writing process for the book was collaborative, and we met throughout the 2018–19 school year to reflect on what Jeremy and Wiline were doing, respectively, in their middle school and university science classrooms. These conversations were full of the enthusiasm that educators can bring while sharing ideas, considering new approaches, and making connections to what works across teaching contexts. As we finalized the book in the late summer of 2019, we were again energized for the upcoming school year.

In previous collaborations, Jeremy and Troy had used a blended voice to reflect their shared writing style, relying on the first-person from Jeremy's perspective to, as we noted in the prefaces to those books, share a teacher's voice from the classroom. Working with three authors, across two classrooms, this book presented more of a challenge.

To that end, you will find spots where we take on the collective "we," and those sections are authored collaboratively. In other places, we will clearly identify the author who is in the lead at that moment, often employing the first-person. We hope that the net result of our work is still a smooth read, and that educators from various contexts can find the book useful in many ways.

We, of course, enjoyed the process of working together and are looking forward to our fifth summer institute at Beaver Island in 2020, continuing to refine and adapt the lessons that we learn from our own students as we share them with other engaged, impassioned educators.

A Note About the Companion Website

Throughout the book, we will have a multitude of resources to help you along the journey of science and literacy. We have created a companion website where you can find these resources. Visit **https://jeremyhyler40.com/science-and-literacy/** to start finding resources. As you continue through the book, there are many times we reference our site. Please feel free to visit the site at any time to access the materials you find useful.

Before getting to the rest of the book, more than a few acknowledgments are in order:

- First, the team from Eye on Education/Taylor and Francis as well as Swales & Willis, including Lauren Davis, Publisher, and Karen Adler, Editor, have been tremendously supportive. Additionally, as we moved into the final stages of production, we appreciated help from Lucy Stewart, Editorial Assistant, George Warburton, Production Manager, and Emma Lockley, Copy Editor.
- Second, we offer thanks to all the teachers who have attended the four years of our Beaver Island institute, and specifically to the collaborators with whom we plan and lead the institute. Jeremy Winsor, Merideth Garcia, and Courtney Kurncz have all been incredible thinking partners, allowing us to share many of the ideas in this book during the institute, all the while providing much needed feedback on our strategies.
- Third, the Beaver Island Institute work has involved a number of people from Central Michigan University (CMU) as well as external funding, allowing us to bring a total of about 50 educators to the island over four years. This is a case where there are too many people to name, and in doing so we would forget someone, so we offer a hearty thanks to everyone at CMU who has, in ways small and large, made this institute possible for the past four years and, we hope, for many years to come.
- Finally, we do add one specific name here at the end. Our thanks to CMU doctoral student Michelle Claypool for her careful proofreading of the manuscript.

Sincerely,
TH, JH, and WP

Link for the Companion Website

To make your professional learning experience with this book, we hope, just a bit easier, we've provided all the links to online resources mentioned within the text on a companion website.

Our book's page is hosted on Jeremy's domain, and you can find the main site with links to chapter sub-pages at https://jeremyhyler40.com/science-and-literacy/, by using this alternative short link <http://bit.ly/askexplorewrite>, or by scanning the QR Code below.

Meet the Authors

Troy Hicks

Dr. Troy Hicks is Professor of English and Education at Central Michigan University (CMU). He directs both the Chippewa River Writing Project and the Master of Arts in Learning, Design and Technology program. A former middle school teacher, he collaborates with K–12 colleagues and explores how they implement newer literacies in their classrooms. In 2011, he was honored with CMU's Provost's Award for junior faculty who demonstrate outstanding achievement in research and creative activity, in 2014 he received the Conference on English Education's Richard A. Meade Award for scholarship in English Education, and, in 2018, he received the Michigan Reading Association's Teacher Educator Award. Dr. Hicks has authored and co-authored numerous books, articles, chapters, blog posts, and other resources broadly related to the teaching of literacy in our digital age. Follow him on Twitter: @hickstro

Jeremy Hyler

Jeremy Hyler is a middle school English and science teacher at Fulton Schools in Middleton, Michigan. He is a teacher consultant for the Chippewa River Writing Project, and a 2019–2020 Community Ambassador for NCTE. Hyler has co-authored *Create, Compose, Connect! Reading, Writing, and Learning with Digital Tools* (Routledge/Eye on Education, 2014) with Troy Hicks as well as *From Texting to Teaching: Grammar Instruction in a Digital Age (2017)*. Jeremy is also a MiddleWeb blogger. His column is called 'Create, Compose, Connect'. He can be found on Twitter @jeremybballer and at his website jeremyhyler40.com

Wiline Pangle

Dr. Wiline Pangle is a faculty member in the Biology department at Central Michigan University. She teaches a variety of classes, which include entering freshman introductory Biology courses, upper-level courses on Evolution and Behavior, and graduate level writing courses. Her research is centered on the behavior of mammals, especially the decision-making processes involved in antipredator behaviors. She has received prestigious grants and fellowships, including the American Association of University Women International Fellowship, for her doctoral work carried out in Kenya on spotted hyena behavior. Wiline also explores the intersections of science and the performing arts, with dance, music, and costume designer professors. Her recent collaborative work has been featured on Michigan Public Radio and in *Forbes* Magazine.

Introduction: Our Science/Literacy Stories

For anyone who has been involved in teaching language arts or science over the past decade, there have been more than a few shifts in curriculum, instruction, and assessment. As with many aspects of life, when it comes to teaching science — and integrating literacy skills into the teaching of science — the only constant is change.

Of course, there are the Common Core State Standards (CCSS) in English language arts, ELA, (released in 2010) and the Next Generation Science Standards, or NGSS (released in Achieve Inc., 2013a). These two curricular documents — created in large part through the work of Achieve, an educational organization "[c]reated in 1996 by a bipartisan group of governors and business leaders" and designed "to help states make college and career readiness a priority for all students," (Achieve, Inc., 2018) — have shaped the past decade of reform. Additionally, in 2016, the International Society for Technology in Education (ISTE) refreshed its student standards, and they are beginning to permeate curricular conversations as well (International Society for Technology in Education, 2016).

At the same time these standards have been updated and released, there has also been a renewed focus on subjects such as science, technology, engineering, and math. While it was difficult for us to pin down an exact date when the term "STEM" was used in the research literature or mentioned in education-related journalism, a quick query into a major search engine yielded over 326 million hits for "STEM Education" as of the time we finalized this manuscript in September 2019. Suffice to say, the world of STEM (as well as associated movements such as "STEAM," which includes Art, and "STREAM," which could either include Reading or Recreation, depending on who you talk to) continues to grow.

And, when there is an ever-widening discussion about the importance of writing across all disciplines, we are caught up in a moment where teaching literacy and science in tandem are becoming more and more the norm. To further that point, a 2017 blog post by educator Maryellen deLacy summarized many of the connections in the following manner:

> While that goal is one set by Common Core ELA standards and literacy has traditionally been the domain of ELA teachers, literacy skills are the foundation of all learning. A **coordinated effort at teaching,**

strengthening, and reinforcing these skills across multiple subjects, including STEM learning, is a practice that makes good sense for both students and teachers.

(2017, emphasis in original)

In addition to being first outlined in the CCSS through the "Grades 6–12 Literacy in History/Social Studies, Science, & Technical Subjects," many of these ideas were described in more detail in the NGSS's Appendix M, "Connections to the Common Core State Standards for Literacy in Science and Technical Subjects" (2013b). Here, the NGSS writing team begins with the point that "[l]iteracy skills are critical to building knowledge in science," and go on to create a cross-walk of CSSS ELA and NGSS standards. Science and literacy, if they were not explicitly connected in American educational policy and curriculum before, are certainly now intertwined.

Still, even with all that has been going on over the past decade, we still have a great deal to learn. Whether we teach in K-12 schools (like Jeremy, at the middle school level), or at the university (like Wiline, in biology, and Troy, in teacher education), we know that blending the teaching of writing into the teaching of science remains a challenging task. There are a number of reasons for this, not the least of which is limited time in our crowded curricula as well as few opportunities for substantive professional learning, which might help us re-envision the role of literacy in science. Our goal then, in this book, is to explore at least a few ways in which the three of us are considering these new challenges.

To begin, this introduction allows each of us to provide some background about how we arrived at the point we find ourselves today. Throughout the book, you as the reader will find that there are points where we speak together in a collective "we," and there are points where we move to our individual voices. We try to make this clear throughout and, with the efforts of the editorial team from Routledge and Eye on Education, we feel we have done so. However, we apologize right up front if there are still any points of confusion.

That said, we use our individual voices in the next three sections of this introduction to describe, briefly, the ways in which we have come to understand the role of writing in science. As a way to begin, our brief biographies are as follows:

- Jeremy Hyler, a middle school language arts and science teacher for 16 years, teaching in rural mid-Michigan.
- Wiline Pangle, a faculty member in biology for 10 years, teaching at Central Michigan University.
- Troy Hicks, a faculty member in English and Education for 12 years, as well as a middle school teacher before that, also teaching at Central Michigan University.

As noted in the *Preface,* the three of us have worked together for four years on a combined science/ELA workshop at Central Michigan University's Beaver Island Biological Station. We call it, appropriately, the Beaver Island Institute (and a link to the institute homepage can be found on the book's companion page for this chapter: https://jeremyhyler40.com/science-and-literacy/introduction). Our stories of literacy learning in science are different and, like your own students' stories, can provide some insight into the ways that we understand what it means to be a reader and writer while also being a scientist.

We share each of our stories below. Then, at the conclusion of this chapter, we provide a few other shared thoughts about the role of writing in science as we head into the main section of the book. First, let's meet Jeremy.

Jeremy's Story: My Path to Becoming a Mad Scientist...err, Science Teacher

As I am learning more and more about teaching science, as with anything in life, I am becoming more and more confident in the work that I do. Still, I make mistakes, and I keep learning. This book is one more step in my journey.

I started my teaching career in a self-contained elementary classroom where science was just one ball of many that had to be juggled. I have never really thought of myself as "just" a science teacher, or someone who would have a love for teaching science. After all, the only science lessons I had put together were the ones from my methods class while completing my undergraduate degree. Still, I forged on for four years in the early 2000s, in my self-contained classroom, doing the best that I could with science.

It wasn't until 2006 that I started teaching middle school, and suddenly science was one of my main subjects that I was teaching. My daily schedule consisted of mostly English language arts (ELA); science was one hour a day. And, to be honest, I liked it that way. I wasn't confident in my skills as a science teacher and I really had no interest at getting better. Reading the teacher's manual was helpful and gave me a basic idea of what I needed to do each day. However, that was about it. Doing a science lab with my students almost never happened, and oftentimes we were watching videos on the content we were covering. No surprise, I wasn't the most confident in teaching science. In retrospect, I am sure I did an adequate job, but I know I could have done better.

Eventually, the day came where I, alone, became the entire ELA department in our middle school. Our district was going through some changes and my teaching assignment changed to where I was just teaching ELA: seventh and eighth grade. Science was out of the picture, for a while. I would be lying if I didn't say I was happy about the new transition because, well, I felt like I was struggling to teach science. So, from 2009 to 2016, I was not directly teaching science, and that was OK with me.

It was also during this time that I started working with my colleague Jeremy Winsor on cross-curricular projects, specifically the Salmon in the Classroom Project (SIC) sponsored by the Michigan Department of Natural Resources (DNR). Together, we "Jeremys" collaborated so students would produce cross-curricular content in science, social studies, math, and ELA. For my class, students wrote an argument piece about the benefits of the project for our ecosystem and economy. Though I was connected to science, I did not directly have to teach science. And, as noted above, that was still just fine with me.

It was also during that time Winsor (my affectionate way of naming my friend and not sounding like I am talking about myself in the third person!) and I were invited to Beaver Island, Michigan (2016) for a week long science/literacy institute sponsored by Central Michigan University's STEM Education program and the Chippewa River Writing Project. Having been part of the Chippewa River Writing Project for many years at that point, I knew that the event would be good, and convinced Winsor to come along. And, well, I knew we would get to hang out with Wiline and Troy, which was a gift in itself.

We were exposed to plenty of field work in science and learned some valuable literacy skills to apply to our own classrooms. In addition, we were asked to complete a project that involved science and literacy that we would use in our classroom. Winsor and I took that opportunity to align more of our science and ELA curriculum (especially with the release of NGSS) and we refined our SIC project. It was only then, during that time on the island that I started to develop a greater appreciation for science and to fully understand that science and literacy can work together so well. The journey had been long and arduous, but I began to gain confidence.

Also, in the summer of 2017, Winsor and I were invited back to the Beaver Island Institute as co-facilitators. With Winsor being the exceptional science teacher he is — and with my knowledge of tech in the ELA classroom — we were tapped to help lead other teachers become more aware of ways to weave literacy into science instruction. It was during that year I would take the plunge back into the science classroom.

For the next two years (2016–17; 2017–18), I taught 8th grade science. In this new curriculum, based on the NGSS, we would be focused on human impacts on the environment. In particular, we would be studying water quality, run-off, pollution, and, a topic especially important here in Michigan, polyfluoroalkyl substances or PFAS (State of Michigan, n.d.).

In the summer of 2017, Winsor had also been through a three-week workshop from the American Modeling Teachers Association (AMTA) on an instructional technique called "modeling" and I was able to see firsthand how that was going to work in a high school classroom (American Modeling Teachers Association, n.d.; Jackson, Dukerich, & Hestenes, 2008). Oftentimes, Winsor and I simply call it "whiteboarding" (which we will elaborate on in Chapter 4). This new instructional strategy was going to be ideal as it was going to require more inquiry in my classroom; it wasn't the "sit and get" method that I had practiced in the past or what students want us to do as teachers. In addition, the science department organized the curriculum to better suit the students getting ready for high school, making it easier for me to see what I had to get through to prepare my students.

In addition to whiteboarding and new curriculum expectations, I also lead the 8th graders through the SIC Project. Needless to say, I was a bit nervous about it, but I had great support from Winsor, who has been doing SIC for nine years. As I concluded, my second year as the 8th grade science teacher, I felt more confident in what I was doing overall. I had a grasp on the curriculum, and I had become more organized in how I wanted my year to look with the students.

Then, another curveball. For the 2018–2019 school year I found out that I was going to be teaching 7th grade science. As a matter of fact, I was not going to be teaching any 8th graders in any subject. Though I was just starting to get my feet under me, change was upon me again; this time I had to pull up my teacher bootstraps and figure out what was going to happen in my 7th grade classroom. Suddenly, I was facing plants and photosynthesis, organisms on Earth, non-living Earth systems, the Earth's surface, the relationships between living and non-living things and, for good measure, a little astronomy too.

While I thought about teaching all of the new topics, once again, Winsor and I found ourselves headed back to Beaver Island as co-facilitators, and Winsor had participated in (and facilitated) more AMTA training in whiteboarding. I watched him and I was able to pick up more of what needed to be done with students as I went through a lab/lesson with modeling instruction. Also, as I watched other teachers go through it, I witnessed the discussions that took place and I was feeling really confident in this method of teaching. While on the island, Winsor and I were able to have more in-depth conversations about having our school adopt this modeling practice. Part of what I am doing then, through the writing of this book, is to document the use of modeling and whiteboarding in grades 6–12, as it happened throughout the 2018–2019 school year.

Whiteboarding has been an awesome experience to do as a department, and I was feeling really good about teaching 7th grade science. In fact, it was the best I had felt about teaching science ever in my entire career. I knew I had a solid curriculum, a great instructional approach, and more importantly I had colleagues who were willing to work hard, be supportive, and do what is best for all of our students.

So, I have invited you to take the journey with me to see how my science classroom is not *just* a science classroom, but instead is a science and literacy classroom where students are engaged every day in inquiry, reading, and writing. In front of the classroom stands a confident teacher who has been through quite a journey to get where I am at today. And, I admit that I am still learning. That's the way teaching goes, and I am happy for having the opportunity to be a lifelong learner.

Wiline's Story: from France to Michigan, with Many Stops in the Field

My connections to science and writing are highly intertwined.

To begin, English is not my first language: I was born in France and pursued my education there until the end of high school. I quickly learned that to succeed in the French classroom, I had to be a proficient writer, as all exams are short answers built around argumentative writing. I remember the pleasure I took, even as a science student, in writing long, literature-based essays for my philosophy classes.

It was a large shock to my confidence to go from such a skill back to "zero" as soon as I stepped into my English-speaking university courses in North America. I realized that I had the writing skills of an upper-elementary student, where I could describe my ski vacation in great detail, but was incapable of writing a concise scientific argument. With the perspective I now have, I realize that the situation forced me to learn how to write in science, essentially, from scratch: I had no preconceived notion, no pride, and, as such, no baggage. This was an advantage for me, but it is contrary to what I see with most of my current students.

As a biology student, I learned my English skills at the same time as I progressed through my courses, and I cannot differentiate learning science from learning to write in my second language. This became particularly clear when I was asked to write a summary of a scientific article that would be published for a lay-audience: this proved such a challenge that I remember it to this day as one of the hardest pieces of writing I've had to produce. I am proud to say that, ultimately, it was published (under my maiden name: Trouilloud, Delisle, & Kramer, 2004), and I have continued to have success with academic writing in the years since.

As I started my doctoral dissertation, the importance of writing in science took center stage. I became quickly aware that, as an international student, I did not qualify to apply to most grants to support my doctoral research. There were only a handful of highly competitive scholarships or fellowships available, which were obtained through lengthy applications. The core of the applications consisted of short essays of one to two pages, in which I had to convince readers to support my entire dissertation research. I made it a personal challenge to obtain funding despite the odds against me, and worked diligently, draft after draft, on revising my science writing.

In the moment, this all felt terribly challenging. Yet, with the perspective I now have, this was an ideal situation to learn how to write effectively in science; I was constantly revising my writing, seeking peer-reviews from established professors, condensing long pages into single paragraphs. And yes, I did obtain some funding, and all of this without much formal training. It was when we started writing this book that I recalled how my doctoral studies did not contain a single writing course until the very end of my dissertation. This seems particularly surprising to me now, as I reflect on how the most successful scientists are those that can express their thoughts in writing in a convincing manner.

As I oriented myself more and more toward teaching, I became quickly aware that writing was not an integral part of most science courses. The reason why became apparent the first time I had to grade 100 midterms that were all short-answer questions! While I would argue that most professors with whom I've worked with realize the value of writing (especially when we assess our students), most have also opted for multiple choice tests to avoid the endless grading. The lack of writing in science classrooms, in turn, creates college students who generally lack skills in basic science writing, which they often do not get exposed to until upper-level classes during their senior year.

I have continued to embed writing into all my classes, regardless of whether I am teaching an introduction Biology class to non-majors or a graduate level course. I have realized that I simply cannot teach science without integrating writing, despite the challenge of grading. For instance, my freshmen students write argumentative essays about evolution by natural selection, while my graduating seniors are reviewing the scientific literature and composing term papers presenting their view of a specific topic or hypothesis. Though, sometimes, students might not be particularly happy with me during the course itself, I've received many emails from past students expressing their gratefulness for having forced them to become stronger science writers. Thus, my many journeys — as a second language writer, as a scientist, and as an educator — all converge here, collecting some of my ideas with Jeremy and Troy in the writing of this book, and sharing them with a wider audience of educators, including you.

Troy's Story: Stepping Into Science Literacy

For me, the connections between science and literacy were, fortunately, introduced early in my career, and I have carried them forward over two decades of K-12 and university teaching.

As an English major with an environmental science minor at Michigan State University, I was lucky enough to take undergraduate coursework in literature and writing, as well as biology and environmental studies. One of the first professional conferences I attended featured Harvey "Smokey" Daniels, talking about content area writing, and my senior thesis focused on nature writing in the high school classroom. Even in my student teaching, I had the opportunity to be part of an AP environmental science course. However, my first job was focused on teaching middle school language arts, and that is where my professional interests quickly moved.

Still, as I began to learn more about the teaching of English, in general, and writing, in particular, I found more and more connections to "writing across the curriculum" or "writing in the content areas." My student teaching mentor invited me to be part of a Michigan Department of Education committee, the Content Literacy Committee, and I soon found myself in monthly meetings with teachers from around the state, all interested in figuring out ways to bring writing into math, social studies, and science classrooms. This led to presentations at various workshops and conferences, interacting with other educators who were also interested in bringing writing into their content area classrooms. Additionally, working with my colleagues at the middle school in a grade level team, we were able to talk about some ideas for integrating reading and writing in an interdisciplinary manner.

This work quickly began to overlap with connections in the Michigan Reading Association, and one particular project led us to create "writing-to-learn" handbooks that we published and presented on at multiple conferences across the state. My initial introduction to ideas like "Cubing" (an inquiry strategy moving through six types of questions) and "RAFTS" (inviting students to consider role, audience, format, topic, and a strong verb when writing from a different perspective) were enhanced by my interest in writing and technology. These initial handbooks have, unfortunately, been lost in a sea of digital debris and boxes full of binders in my basement, so it is a challenge for me to cite them here in an academically appropriate manner.

Still, in the early to mid-2000s, once I entered graduate school, again at Michigan State, I had the good fortune of connecting with a local site of the National Writing Project (NWP), the Red Cedar Writing Project. I became quite interested in writing across the curriculum and I was engaging with professional books like Daniels et al's *Content-Area Writing: Every Teacher's Guide* (2007) and *Subjects Matter: Every Teacher's Guide to Content-Area Reading*

(2004), among many others, to discover more strategies. Writing across the curriculum, writing in the disciplines, writing-to-learn; whatever we call it, these are all part of what I understand about what it means to teach writing.

It was in the late 2000s that Shanahan and Shanahan shared their work on disciplinary literacy (2008), and with my new network of colleagues in the NWP, my interest in "disciplinary literacy" continued to grow. Their work, in particular, pointed out a stark problem:

> The high-level skills and abilities embedded in these disciplinary or technical uses of literacy are probably not particularly easy to learn, since they are not likely to have many parallels in oral language use, and they have to be applied to difficult texts… But something else makes these high-level skills very difficult to learn: They are rarely taught.
>
> (45)

The Shanahan's work reiterated what we understand about how the best teachers do their work: they make expert knowledge available to students in a clear way, helping them understand the nuances of being a reader and writer within a discipline (further work on this has been done in Lent's *This is Disciplinary Literacy* (Lent, 2015)). Seeing the many ways in which writing was taught across the university (including the specialized discourses and ways of thinking that these varied disciplines required) all intrigued me, and ever since becoming a faculty member at Central Michigan University and becoming the director of our own Chippewa River Writing Project, I have continued to be enthralled by opportunities for exploring writing in the disciplines.

While teaching pre-service teachers, I was invited to review and pilot a new version of the writing-to-learn handbooks that had then been produced by the Michigan Department of Education, including one on science. One of my most recent projects has been to help develop a "writing intensive faculty workshop," a series of instructional modules and videos that are available to other Central Michigan University instructors, and also open to anyone (for this, and other URLs, please visit our book's companion page: jeremyhyler40.com/science-and-literacy/).

Collaborating with faculty across the university in order to create engaging, useful professional development videos for our colleagues gave me even more insight into the complexity and wonder of disciplinary writing. I've been fortunate enough to continue my own scholarly writing related to disciplinary literacy (Hicks, Bruner, & Kaya, 2017; Hicks & Steffel, 2012) as well as to lead our annual Beaver Island Institute. To discover more about the design and structure of the institute, please see Chapter 6, "Professional Development in Science and Writing."

In working on this book with Jeremy and Wiline, I've continued this line of thinking and inquiry. Through our many conversations, visits to Jeremy's classroom, and our planning and delivery of the annual institute, I've come to learn even more about the role of argument in scientific writing, the ways in which the Next Generation Science Standards are driven by inquiry, and how we can invite students to use digital writing tools to create even more compelling compositions. I know that I will continue to be interested in the intersections of science and writing because the possibilities are, indeed, endless, and we are pleased to share our thinking in this book.

Next Steps

As we move toward our first chapter, we are reminded that our own professional learning and growth never ends. We have approached the writing of this book in much the same way that we have approached the design of the Beaver Island Institute, with a spirit of inquiry and hope that we can use writing as a tool to both discover and share knowledge. The book that we used this past year during the institute was Ward Hoffer's *Science as Thinking* (2009). In the opening pages of the book, she argues that "[t]he goal of inquiry-based science instruction is for learners to develop their understanding of science content, as well as the ways in which scientists think about and understand the world" (8).

To extend this idea, we believe that students can use reading and writing in science classes as a way to understand the world. Like Ward Hoffer, Daniels, and many others before us, we continue this line of reasoning, and we hope to add a few new ideas along the way. In the subsequent chapters, we explore the following:

- In Chapter 1, "Getting Started with Writing in Science," we explore our own histories with writing and science in more detail, and we explain what we see as four key shifts driving the science/writing conversation;
- In Chapter 2, "Science Notebooks," we look at the way that both Jeremy and Wiline integrate these tools in their teaching, examining some student examples along the way;
- In Chapter 3, "Visual Explanations with Infographics," we take a deeper dive into the design decisions that go into these aesthetically engaging texts, considering the ethical aspects of representing data as well;

- In Chapter 4, "Modeling/Whiteboarding," we explore the emerging practice of scientific modeling by inviting students to create whiteboard drawings, representing data in a variety of forms;
- In Chapter 5, "Additional Strategies," we share a number of other strategies that we have used, including a lesson on creative nonfiction writing, expanding disciplinary vocabulary, crafting scientific arguments, and briefly reviewing a number of other apps and websites;
- In Chapter 6, "Professional Development in Science and Writing," we describe the design decisions that we have made in structuring our Beaver Island Institute, sharing some insights that we have learned over our three years leading this professional learning experience; and
- In the conclusion, we reflect on the process of writing this book and consider the next steps for teaching science and literacy.

Again, we each come to the work of science and literacy from different perspectives, with unique experiences that give us insights into when, why, and how reading and writing can work. Moreover, we conceive of "reading and writing" through a broader, multimodal lens (which will become evident as we move through the examples later on in the book).

As we move into our first chapter, taking an even closer look at the role of writing in science, we end with an idea that Jessica Fries-Gaither introduces in her book, *Notable Notebooks: Scientists and Their Writings* (2017). Aimed at elementary-aged children, she writes the book in verse, describing how scientists have used notebooks to document their thinking and dreaming, as well as their questions. In the opening sentences, she makes the point that:

Of all a scientist's tools,
objects rare and common,
the lowly science notebook
is most easily forgotten.

In whatever ways you inspire your students to continue their inquiry in science, we hope that you help them see the power and possibilities of literacy. While we might not have the next Curie or DaVinci in our classrooms, we do have the opportunity to help all students ask, explore, and write, just like scientists do, too. Notebooks, combined with the additional strategies for writing (both in text and multimodally, through graphs and infographics), can be powerful tools for helping our students as they ask their own inquiry questions, and explore the answers through systematic data collection. To understand how, we begin with a brief history of writing in science, at least as it typically happens (or, sadly, may not happen) in school.

1

Writing in Science?(!)

While it probably should come as little surprise given the discrete nature of school subjects (English, science, math, and so on), in educational terms, it seems as though science and literacy are often thought of as very separate entities. When phrases like "all teachers are teachers of reading" started entering the popular conversation about literacy in the content areas, and new ideas like "writing across the curriculum" or "writing in the disciplines" came into fashion in the 1980s and 1990s, we should have all stopped to catch our collective breath.

Even saying this sounds a bit silly, but it bears stating clearly here: we begin this book with the assumption that writing is an integral part of science, and that we only understand the world, and the discipline of science, through language. Somehow, this still seems like a radical concept, as a story from Jeremy's early teaching career, shared later in this chapter, illustrates. Yet, it seems as though there should be no question: asking questions, exploring data, and writing about it are as integral to science as beakers, microscopes, and safety glasses.

Suffice to say, scientists have been writing about their experiments and findings — and reading the writing of other scientists — for hundreds of years. In today's academic landscape, literally thousands of peer-reviewed journal articles are published each year. And, more and more, other outlets for science writing, such as blogs and social media accounts have come to exist, creating more accessible opportunities for the general public than academic journals.

Still, we don't need to look too far back into education's history to see that science and writing have been thought of as separate (and, sadly, this still holds true even in some classrooms today). Though writing has been part of science for as long as we have had science, somehow schools have found a way to keep the two apart until recently. Rather than look at this as bad news, however, we see it as a place to enter the conversation about writing and science. In fact, as we began our collaboration on this project, we realized that it would be helpful to have a bit of background on the alphabet soup that surrounds writing in science: WAC, WID, WTL, WTE (see Sidebar 1.1 for a description of these acronyms). Science and writing are connected, of course. But if your experience in K-12 schooling was like ours, you probably can't be held at fault if you didn't know that.

Thus, this chapter aims to remove the question mark that is currently in parentheses in its title and, in its place, put in an exclamation point. Instead of *Writing and science, huh?* we need to shift to *Writing and science, yeah!*

Not only is there room for writing in science, but it is absolutely essential as a tool for our students to discover ideas, to deepen connections, and defend what it is they have come to know through inquiry. If nothing else, we've tried to capture that idea in the title of the book: ask, explore, write.

So, understanding a little bit about how we got to this point — where we are, in contemporary classrooms, consciously making an effort to integrate writing in the science classroom — will help frame the conversation about where we are going in the rest of this book, providing some background for why Jeremy and Wiline make the pedagogical decisions that they do when integrating science and literacy.

Jeremy's Journey: My Introduction to (Not) Writing across the Curriculum

In my second year in the middle school classroom, it was the week before school would begin, and I was sitting in my social studies colleague's classroom waiting to hear about the district's new focus on writing across the curriculum, what it meant, and what was going to happen with it.

The mood of the room was, to say the least, very sour.

There was the tapping of pencils, and there was the waiting for opening comments from our administrator. Word on the street was already out that writing was going to be required in other content areas because students' writing scores on standardized tests were not proficient. At the time, I was teaching 8th grade language arts and science. There were two other teachers teaching science at that time as well, but they — as well as the other math and social studies teachers in the room — had their lips pursed.

Our administrator broke the news by saying, simply, more writing had to be done in all disciplines. That was it. And, upon hearing this statement, as well as the accompanying groans and sighs, I could guess that some of my colleagues were not interested in doing more work. Besides, I would come to learn something else: they were not confident in their own writing skills, a point that would become clear very quickly.

The English teachers, on the other hand, had their brains spinning in many directions and greeted the news with smiles. We were looking forward to helping with vocabulary instruction, looking closely at informational text and reading strategies, or perhaps even other aspects of "writing across the curriculum" or "writing-to-learn." It's worth clarifying these terms, as we will be returning to them throughout the book. And, please know that these ideas weren't necessarily new at the time, but they were "new to me." At any rate, I was coming to understand the kinds of writing that could be done in other classrooms, including science, as falling into three general areas. Wiline, Troy and I summarize them briefly in Sidebar 1.1.

So, back to the challenge at hand for my colleagues and I in 2006. When it came to using writing in science, the other language arts teachers and I weren't sure what steps we needed to take to help the other content area teachers. This was a moment where curriculum was going to be changing, but we didn't quite know it at the time; the Common Core wasn't on the scene yet, and there was no hint of Next Generation Science Standards. Still,

Sidebar 1.1 A Brief Overview of WAC, WTL, WTE, and WID Movements in Education

> As a big picture concept, "Writing across the curriculum" (or WAC) was being used as a broad category that considers how teachers can use writing to support learning from five-minute in-class prompts all the way through to semester-long projects. It had really picked up steam in the 1980s as colleges and universities, and then K-12 schools, saw a need for students to be writers in all subjects.
>
> Then, within that broad definition of WAC, there seemed to be two sub-categories. First, there was "writing-to-learn" (WTL). In these kinds of tasks, writers are invited to think about subject area material in different ways by using writing as a tool to explore. Usually informal techniques are used as formative assessments, and these tasks can be as short as just a few minutes in class or can entail a longer stretch of writing outside of class. These are sometimes referred to as "content area writing" strategies, too, and we have heard, even more recently, that these have been described as "writing-to-engage" (WTE), too.
>
> Second, there was "writing in the disciplines" (WID) or, as more recently defined, "disciplinary literacy." With these kinds of tasks, writers are asked to compose types of texts that disciplinary experts would; scientists create lab reports, data visualizations, and articles for a wider public.
>
> To summarize an entire field of scholarship in just a few more brief words, while WTL activities can certainly lead to deeper engagements with scientific literacy, producing longer, more substantive texts, WID for science has distinct characteristics including a stricter style, requirements for presentation of data, and adherence to the norms of academic argument. Therefore, WID pieces usually take much longer to compose.
>
> Still, writing across the curriculum (WAC), writing-to-learn (WTL), writing-to-engage (WTE), and writing in the disciplines (WID) all share a common goal: to help students learn with and through writing.

it was exciting to consider. If we could get students writing more throughout the school day, that would be a win for everyone.

Then, reality took over. The tone of our staff meeting changed very quickly, as one of my math colleagues was not so receptive. "I didn't go to college to be a writing teacher, I went so I can teach math," he pointed out. Then, he summed up his feelings, with more than a hint of middle school sarcasm in his voice: "Why should I do this?"

His statement continues to echo in my ears to this day.

Being so early in my teaching career, I hung my head and waited for my principal to reprimand him. That, of course, never came (but, that's a different story). And, like many educational reforms, there were (and in many places, still are) good intentions with writing across the curriculum, but I wasn't sure if there was going to be the follow-through that most initiatives need. As any educator can tell you, most initiatives lack buy-in, resources, time, and money. I could already see that this particular initiative, though it had a clear rationale, was already missing many of these features. Furthermore, I wasn't counting on my district (or our state) to put money aside for this reform. So, in some sense, I couldn't blame my colleague, though his negativity was deflating, and quickly infected the room.

Sitting there, still in awe of what was said, I can remember thinking, "Wow ... we are never going to make our students better writers in our district with such a negative mentality." Moreover, while there was going to be a push for writing across the curriculum, there was no direction or focus. The directive was to figure out what we were going to do to have students write more across content areas.

That was it.

There was no support, and in a matter of 45 minutes (well, maybe just four or five minutes, really), it was clear that no one wanted to have students write more in their classrooms. There are many writing avenues that could have been explored when thinking about the implementation of cross-curricular work — different genres, audiences, and purposes — but at that moment, in that room, there were no clear options on the table. No discussion of writing in science, or any other content area, was to be had that day.

I was under the assumption (which ultimately came to pass) that we were all going to go and do our own thing when it came to writing in other content areas. So, over the next few weeks and through the remainder of the school year, I chose a writing assignment that I would collect each marking period to put in my principal's hand. I did so, diligently, asking my students to write and gathering their responses. Needless to say, those papers were never seen again, let alone discussed in department meetings or full staff meetings.

Having students do simple writing tasks without a purpose was not going to make them better writers. And, asking teachers to collect those papers without knowing the purpose for how the writing was going to be used was just deemed as "one more thing to do." My thinking had to change; the writing I collected from my students should be a formative assessment, a chance to show where my students were at with their writing skills at that moment as a way to demonstrate growth, regardless of what my principal or colleagues might ever do with their writing.

So, when I started teaching science again in 2016, I hadn't given it much of a thought in eight years (sad, I know, but I know that many other teachers can relate to this kind of reassignment, as I shared earlier). There were many things that were very new to me where I had to get my mind wrapped around what I was doing. There were a few years where I was focused only on ELA, but I'm back in science now, much more confident in teaching the content.

However, inserting more deliberate moves with literacy and writing into the science classroom is a learning process. My intentions were always good, but when there are curriculum shifts, such as the NGSS — ones that demand that students "think critically, analyze information, and solve complex problems" (Achieve, Inc., n.d.) — and the feeling of, basically, being a newbie again, there were bound to be mistakes. However, my passion to teach science was there, and I had very knowledgeable colleagues who would help along the way.

So, as I have with my previous two books with Troy — *Create, Compose, Connect!* (2014) and *From Texting to Teaching* (2017) — I wanted to make my thinking about all of this very transparent. To that end, this book documents another year in my life, this time as a (re)new(ed) science teacher, with help from my colleague at school mentioned in the Preface, Jeremy Winsor, as well as my co-authors, Wiline and Troy.

The shift into new curriculum was not completely unfamiliar to me because, when Troy and I wrote *Create, Compose, Connect!*, I had transitioned into using the Common Core State Standards (CCSS) in ELA. In that book, we documented the breakdown of the three main genres of writing: narrative, expository, and argument writing. We also discussed the deliberate ways technology could be used in a Language Arts classroom, which is required by the CCSS. Finally, we outlined ways to incorporate listening and speaking lessons into the Language Arts classroom. Similarly, in *From Texting to Teaching*, we thought again about how to layer in technology with the teaching of grammar. While we didn't specifically address ISTE standards in that book, we were certainly influenced by the work of Liz Kolb and her Triple E framework (2017), and that type of thinking will show up from time to time in this book, too.

Now that my head was wrapped around the Common Core and thinking even more intentionally about the use of technology — and, of course, because I am back in the science classroom again — it is time to think about the NGSS. While I'm still learning, here are a few things that I've noticed about the NGSS, and we will bring those ideas up throughout the book. For the moment, we move back to a collective voice where Wiline, Troy, and I think about some of the reasons for embracing writing in science and the history of science and writing.

A Bit of Scientific Writing History

The connections between science and writing are deep and intertwined, as the three of us have discussed many times over the past four years in our work on the Beaver Island Institute.

Certainly, we could not produce scientific knowledge without language (or, for that matter, mathematics). Yet, at the same time, scientists struggle with precision, attempting to classify and define organisms, elements, processes, and other scientific phenomena in words, we also know that language is, if anything, imprecise. This is, paradoxically, probably one of the reasons why we have hesitated to teach students about writing in science, though it is clear that we really should be. Writing in science is challenging, and that is precisely why we need to take it on.

It's hard to know exactly when the first scientific writing was produced, though some would argue that it began with the work of the ancient Greeks, especially Aristotle, and certainly by the end of the medieval age with DaVinci (1452–1519), Copernicus (1473–1543), Galileo (1564–1642), and Newton (1643–1727) as very early examples. Over the centuries, scientific writing has become more formal, beginning with the very earliest examples that date to the founding of the Royal Society of London in 1660, to the work of Darwin in the 1800s (much of it in his personal journals). In the twentieth century, we have been introduced to the work of influential writers like Rachel Carson and E.O. Wilson, and the hundreds of thousands of scientists working in labs, research facilities, and universities around the world, right up to the present day. Certainly, we've missed a few great scientific writers in our list here, and our companion website has a collection of some of our favorites, largely curated by Wiline and, as appropriate, added to by Jeremy and Troy.

Despite the many centuries of scientific writing that we can point to — much of it very eloquent, though some certainly filled with some confusing and jargon-filled language, too — we don't always see the kind of high-quality, engaging scientific writing that we would like to see in our K-12 and higher ed science classrooms. In looking at the textbooks we ask our students to read, scientific knowledge seems to be codified, shown as distinct facts that have simply always existed. Separate from the inquiry and interpretation that it has taken to arrive at these distinct ideas, the information in textbooks is, more often than not, free of a great deal of context about its development. What this has led to, as Wiline explains below, is a codified version of science that does little, unfortunately, to help inspire students as learners.

Wiline's Perspective: The Dreaded Science Textbook

Science textbooks are often the only representation of science writing that our students get to experience. And, because of their importance in the scientific understandings that our students develop throughout their educational careers, I have a lot to say about the quality of science textbooks!

In short, science textbooks cover lots of material; in reality, too much material. Nothing strikes me as truer than when looking at introductory biology textbooks that cost as much in dollars as they weigh in pounds. The average biology textbook — and I've probably reviewed or taught from at least two dozen of them in my career — cover so much material that it takes instructors two to three semesters of instruction to cover the gist of the textbook, and that is at neck-breaking speed.

Each cycle of revision for these textbooks typically adds content, as publishers strive to please all users of a book by adding all possible topics. And this content is often critical, as the culture of "teaching the book" remains common in both K-12 and higher ed; in other words, if something is in the book, it becomes part of the course curriculum, often in the exact order of the book's table of contents. This results in a reductive curriculum that serves no one well, especially the students learning to become biologists.

Moreover, science textbooks have often characterized the actual process of *doing science* into the well-known "scientific method." This distilled version of the process is taught starting as early as 1st grade in its famous five-step method (asking a question, making observations, developing hypotheses, testing hypotheses, and making conclusions). Not only is this approach to science far from the reality, it misrepresents the very essence of scientific discoveries as a cookbook-style recipe devoid of creativity, error, collaboration, and meanderings. The inquiry process that is central to science is, unfortunately, lost.

In this sense, textbooks don't tell our students the story about what scientific writing is or could be. Science writing is all about argumentation. In *What Science Is and How It Works*, Gregory N. Derry devotes an entire chapter to logic, evidence, and argumentation. In particular, as a process of working through ideas, he notes that "the quality of the argument rests as much on the evidence offered in support of the argument as it does on the logic" (2002, pp. 92–93). What counts as evidence — and how that evidence is drawn from well-designed experiments and observations — matters depending on the scientific phenomenon under consideration.

Crafting a scientifically-valid argument is both difficult and rewarding intellectual work. This balance of rigor and joy is lost in the banality of textbooks.

Moreover, presenting the process of science without context is problematic. For sake of being concise, science textbooks omit the history behind what are now considered scientific theories (including the sometimes convoluted or controversial aspects of those theories). I use the word "theory" here in its scientific meaning, defined in science as an explanation of a natural phenomenon that has been extensively tested and repeatedly confirmed. This is in contrast to a "hypothesis," a suggestion to attempt to explain a natural phenomenon. By removing the context in which discoveries are made — and later revised — scientific knowledge can be presented in a vacuum, as a fact that only a single scientist once discovered and codified. This vision of science is so removed from the reality of actual inquiry and discovery, both which rely on prior knowledge and extensive collaborations. In Sidebar 1.2, we look at one just example: evolution.

To their credit, the publishers of science textbooks are currently issuing a wave of revisions to better illustrate the process of science. Again, I've reviewed dozens of textbooks over the years, and I am pleased to see some subtle changes. For instance, some biology textbooks now include items such as:

Sidebar 1.2 Building Scientific Knowledge Around Evolution

The importance of grounding a body of scientific knowledge in its history has been shown to be particularly critical for the presentation of potentially controversial topics, such as the theory of evolution by natural selection. Teaching evolution has received extensive attention from education research, and there is a consensus among researchers for the need to place such topics within a context (Jensen & Finley, 1995, 1996, 1997). As such, we now recommend that students should be presented not just the theory of evolution by natural selection, but also:

- the scientists that came before the hypothesis was developed by Darwin and Wallace ("the influences"),
- the thought process that came before the hypothesis (as seen in Darwin's science diaries),
- the reaction of the scientific community to the hypothesis at the time,
- the testing of the hypothesis by hundreds of scientists that followed,
- the lines of evidence that came out in various fields, and
- the consensus to rephrase the hypothesis as a theory, some almost 100 years later (Hermann, 2007).

In this manner, something that used to take just a day or two of class time (introducing the theory of evolution by natural selection) now becomes an entire unit of study. And, to be clear, I (Wiline) think that this is good; taking the time that we need to fully explore a scientific concept and inviting students to engage in inquiry related to that concept are both crucial to good teaching and to good scientific writing.

- Vignettes about the scientists and their experiments to illustrate how a specific concept covered in the chapter was originated;
- Connections to specific data from peer-reviewed scientific publications that are pulled out and presented in a simplified format to illustrate how scientists go from initial data all the way to factual information presented in a science textbook, with original references available;
- Applications to everyday life, where evidence is used to guide decision making in common topics, such as whether we should eat chemically-altered chocolate that melts better in our mouths, or whether we should use bottled or tap water; and
- Extensive online supplemental materials, some of which include tutorials, online virtual labs, and extensions, all aimed at practicing scientific thinking and skills.

So, textbooks are imperfect — in being superficial in coverage — and should be viewed as just one part of the curriculum: a support and an aid. Textbooks, in and of themselves, are not *the* curriculum to follow from start to finish. And as such, textbooks should be used along with many other resources, including more traditional sources such as actual scientific articles, video clips from documentaries, and nonfiction books. Additionally, teachers might consider the resources we have made available in Appendix A and on our book's companion site jeremyhyler40.com/science-and-literacy/.

Given this history of how we think about scientific writing, the science writing that our students read, and the science writing that we ask them to do in our classrooms, let's focus now on the alphabet soup introduced earlier: WAC, WID, WTL, and WTE. In thinking about what it means for our students to write as scientists, there are some helpful principles that we can pull from the history of this movement. In the next segment, Troy introduces us to a little bit of history related to writing across the curriculum and that alphabet soup we introduced earlier.

Troy's Perspective: A Deeper Look at WAC, WID, WTL, and WID Movements

Beginning in earnest in the 1970s, and continuing throughout the 1980s and 1990s, numerous books, articles, conference presentations, and other teaching resources were developed under the broad umbrella of "writing across the curriculum" (or WAC). As described by Bazerman et al, the WAC "movement provided systematic encouragement, institutional support, and educational knowledge to increase the amount and quality of writing occurring in such courses as history, science, mathematics and sociology" (Bazerman, Little, & Bethel, 2005, p. 9). These institutional efforts, most often at universities, though also in some degree in high schools, led to instructors adapting strategies that could be used broadly, "across the curriculum," in any subject area.

For instance, I can remember conversations about writing across the curriculum in professional development and staff meetings from my time as a middle school teacher, and it is certainly a topic of interest that I have followed in my career as a teacher educator. Of the resources that I have shared with others in professional development sessions over the years, Harvey Daniels' work stands out. Using the terms "content area reading" (2004) and "content area writing" (2007), Daniels and his colleagues shared a number of strategies that could be used across a variety of teaching contexts. In *Content-Area Writing*, Daniels, Zemelman, and Steineke characterize writing to learn with the following descriptions: short, spontaneous, exploratory, informal, personal, one draft, unedited, un-graded (22). They reiterate the fact that, unlike more formal pieces of writing, writing-to-learn (WTL) activities are designed, intentionally, to help students articulate their thinking, ask questions, and demonstrate understanding. Writing-to-learn, in combination with more formal writing, provides writers with "balanced opportunities to operate across the whole spectrum of language uses, contexts, and levels" (24).

These approaches were developed as a result of research done on process-oriented pedagogy for reading and writing instruction. The general argument for using such strategies in content-area courses follows this line of thinking: because writing is thinking made visible, it should be possible for educators in any content area to use writing as a tool for students to engage in any number of tasks, ranging from the lower levels of educational objectives (from Bloom's taxonomy) such as memorization and recall all the way to higher levels that require synthesis and evaluation.

And while my own colleagues and others, like Jeremy's math colleague he introduced earlier, might still disagree, research has now shown that writing increases retention and comprehension of scientific knowledge (Klein & Boscolo, 2016). This includes some of Wiline's work (Linton et al., 2014), who examined with colleagues the effects of writing and discussion in class that led to a change in performance on final exams. In their experimental study of students who used WTL activities, they discovered "that students who write about a concept perform better on subsequent writing-based assessments of that concept compared with students who only discuss the concept with peers in cooperative groups" (2014, 19). By engaging students in a variety of writing tasks, ranging from short, in-class assignments all the way through to fully scaffolded, unit-or semester-on projects, instructors from subject areas other than English language arts would, sensibly, be able to introduce their students to material from their courses through the act of writing. And, in recent years, the conversation has continued to move, as a new term has come into the writing across the curriculum conversation: "disciplinary literacy."

A Move Toward Disciplinary Literacy

In 2008, Timothy and Cynthia Shanahan released their seminal article, "Teaching Disciplinary Literacy to Adolescents: Rethinking Content-Area Literacy," in *Harvard Educational Review*. Shanahan and Shanahan (2008) began with the premise that, for decades, everyone involved in literacy education — teachers, literacy coaches, and professors of literacy education — had been approaching comprehension instruction from a faulty model. With the goal of teaching general strategies that students would accumulate over the course of their K-12 career, leading them ultimately to becoming proficient at reading discipline-specific texts in upper grades and in college, educators had hoped that this approach would lead to success in disciplinary contexts. In other words, from this perspective, students should be able to transfer what they learned from generic reading and writing strategy instruction into the technical and contextual language of disciplines. While some students are able to successfully navigate these transitions, many are not.

Through their work, Shanahan and Shanahan explored the ways that experts in a variety of disciplines approached literacy tasks. Rather than looking at generic skills, they were interested in the very specific ways that experts engaged in reading and writing with discipline-specific texts. Shanahan and Shanahan concluded that "disciplinary experts we studied approached reading in very different ways, consonant with the norms and expectations of their particular disciplines" (51) and, because of this, "[s]tudents' text comprehension, we believe, benefits when students learn to approach different texts with different lenses" (51).

Like many aspects of teaching and learning, however, this is easier said than done. In addition to discovering discipline-specific ways in which experts attended to reading, Shanahan and Shanahan suggested that there was another problem — which we first mentioned in the *Introduction* — that "makes these high-level skills very difficult to learn: They are rarely taught" (45). As students' progress into higher grades with deeper demands of discipline-specific vocabulary and knowledge, they argue that "literacy instruction often has evaporated altogether or has degenerated into a reiteration of general reading strategies" (45). In other words, the types of generic, reading- or writing-to-learn strategies that are useful in helping students become literate in their early years are not enough to help them become literate in discipline-specific domains.

This work created opportunities for more in-depth study related to disciplinary literacy that moved beyond strategy instruction alone. In an edited collection, Jetton and Shanahan argued that "[w]hile these strategies are important and effective, they are not enough. Students must also understand the special nature of each academic discipline" (2012, p. x). Almost overnight, it seemed, those of us who had looked at content-area literacy strategies as a primary way to engage students in reading and writing throughout the curriculum were suddenly realizing that a writing across the curriculum or writing-to-learn approach was not enough. Thus, the field of disciplinary literacy exploded.

In the past decade, numerous journal articles, conference presentations, and books have all explored the emerging idea of disciplinary literacy. Connecting to another disciplinary literacy expert, ReLeah Cosset Lent opens up her book on the topic by stating that "reading, writing, thinking, reasoning, and *doing* within each discipline is unique — and leads to the understanding that every field of study creates, communicates, and evaluates knowledge differently" (2015, p. 1, emphasis in original). It is this idea of "doing," the idea that literacy is simply a part of being a scientist (or, for that matter, a mathematician, historian, artist, or other professional), that has perhaps solidified the nascent ideas from the early days of the WAC movement.

So, do we see that all teachers consider themselves to be teachers of literacy? Well, unfortunately, I don't think we're there quite yet. As we write this book, I am a member of a statewide "literacy essentials" working group in Michigan that is working to implement disciplinary reading and writing standards in grades 6–12, and this is after having been part of the committees and projects I noted from earlier in my career. In short, the work continues. Still, things have changed a lot in the past three decades, and if what we see emerging from new content standards — including the Next Generation Science Standards — are any indication, then perhaps we might be getting closer to this goal than we ever have before.

Strategies for Writing in Science

A number of strategies have been presented in a variety of teaching materials, so often that it is difficult to ascertain exactly where some of them originated. Many resources have documented these strategies, including a booklet produced by the Michigan Department of Education, "Writing Across the Curriculum: Science" (Michigan Department of Education & Michigan Science Teachers Association, n.d.) and WritingFix.com's "Writing Across the Curriculum" guide (both linked from our companion website: https://jeremyhyler40.com/science-and-literacy/chapter-1/). At the university level, one of the most popular texts for exploring the possibilities of writing-to-learn is John Bean's *Engaging Ideas: The Professor's Guide to Integrating Writing, Critical Thinking, and Active Learning in the Classroom* (2011). A quick search of the web will yield many more resources, and across all of these resources, a number of themes resonate.

In summary, those who are advocates of writing to learn believe that, depending on the task, time, and overall approach, the process can vary significantly. In terms of time, these activities can range from a brief two–three minute in-class writing break to an extended piece of writing that might happen in or outside of class.

Then, in terms of implementation, this can happen within a single class period, or across class periods. Within a single class period, there are opportunities for including writing as a way to begin class, to prepare students for and then initiate conversations, as a break during class to have students reflect on what they have been learning, or at the end of class as a way to invite student comments and questions. As a pattern of writing across multiple class sessions, these can become routines that make teaching and learning more effective. Finally, these WTL activities vary in terms of formality, with options for personal, informal pieces of writing that may not even be shared with the class or instructor, and are really only for the students' own thinking. Or, the activities could be turned in as a class assignment for a formative assessment. Regardless of implementation method, the point here is that WTL can be used in many ways, and be brought to the fore through many strategies, as shown in Sidebar 1.3.

Given the dozens of different kinds of writing-to-learn strategies that already exist, instructors may wonder why they might need to still attempt to create their own writing tasks for students. There are many responses to this question, but the short answer is this: teachers know their students — and their content — the best, and there is no better person to design a writing

Sidebar 1.3 Brief Summary of Popular Writing-to-Learn Strategies

Just a few of the tried-and-true WTL strategies from the sources mentioned above include:

- Two-column (Cornell) notes, in which students divide their notebook to record notes directly from their teacher's presentation in the left-hand column and, in the right, ask questions, make connections, or draw conclusions about the information being presented.
- Venn diagrams where students create overlapping circles and record information about two ideas on the outsides of the circles and, where they cross, describe similarities between the topics.
- Frayer squares/definition maps that invite students to put a topic in the center and then, in the four quadrants around that topic, explore the definition, characteristics, examples, and non-examples related to the topic.
- RAFTS (Role, Audience, Format, Topic, and Strong Verb) as a creative, yet exploratory way for students to examine a topic from an alternative perspective, describing a particular topic through imagery and personification.
- Exit Slips that students complete at the end of a class session, in which they can share significant ideas that they have learned, questions that remain, or connections to other course content.

assignment than the person most familiar with both. With that in mind, one additional resource that is highly useful is Traci Gardner's *Designing Writing Assignments*, available as a free download courtesy of the National Council of Teachers of English and the Writing Across the Curriculum Clearinghouse at Colorado State University (2011).

Gardner begins with the assertion that designing writing assignments is one of the most important tasks that teachers of writing must engage in. She notes that "Writing assignments are, in many ways, the structure that holds a writing class together" (2011, x). To that end, she offers throughout the book a number of strategies for teachers, as designers, to think critically, carefully, and creatively about the ways that tasks can be developed in order to fully engage students and support their intellectual growth. In particular, Troy has found Chapter 4, "Defining New Tasks for Standard Writing Activities," to be very useful. In this section of the book, she invites teachers to think about five critical questions that they can use in terms of designing the assignment. She asks:

- "Who will read the text? Can I choose an alternative audience?"
- "What stance will students take as writers? Can the assignment ask for an unusual tone?"
- "When does the topic take place? Can the assignment focus on an alternative time frame?"
- "Where will the background information and detail come from? Can the assignment call for alternative research sources?"
- "Can students write something other than a traditional essay? Can the assignment call for alternative genres or publication media?" (2011, 49)

Subsequently, Gardner provides readers with many, many examples for students to use as alternative audiences, stances, and genres to traditional academic essays. Consider the example that, rather than writing a lab report about the states of matter and how water freezes, thaws, and evaporates, we could instead have students create a travelogue, where the drop of water would describe her or his path through these different stages. Alternatively, an op-ed could describe how water can never really be trusted because, in large amounts, it can actually be poisonous to humans (do a search for "dihydrogen monoxide" and enjoy the DMHO.org website, and use it as a resource for teaching critical evaluation skills!). The creative possibilities are quite endless, and Gardner provides plenty of inspiration.

It is with this brief history in mind that we now refocus our attention on the current context for teaching disciplinary literacy skills, especially writing in science, with a deeper dive into the NGSS.

Literacy Now in the NGSS

The NGSS brings a new set of perspectives to science instruction, most notably an inquiry-based approach that will require teachers and students both to think about using writing in ways they may have not yet considered. Without going into a detailed history, the NGSS were developed in an iterative process involving "practicing scientists, including two Nobel laureates, cognitive scientists, science education researchers, and science education standards and policy experts" (Achieve Inc., 2013a). After this team released a draft of their Framework for K-12 Science Education in July of 2011, then the NGSS were developed over a span of two years, and involved "stakeholders in science, science education, higher education, and industry" (Achieve Inc., 2013a).

The entire NGSS process was spearheaded by Achieve, the organization that also led efforts on the CCSS. Depending on who you ask, that could be either a good thing, or a bad one. And, as with the development and implementation of the CCSS, there has been much debate and discussion about the NGSS, though delving into that is far beyond the scope of our book (as one example, see McComas & Nouri, 2016). Both Wiline and Jeremy — who have examined and used the NGSS in curricular planning — have their concerns about the clarity of the writing in the standards and whether they are approachable and practical for K-12 teachers, as they implement the ideas in their own classrooms.

That said, we do appreciate the ways in which the NGSS addresses disciplinary literacy. As noted in Appendix M of the NGSS:

> As the CCSS affirms, reading in science requires an appreciation of the norms and conventions of the discipline of science, including understanding the nature of evidence used, an attention to precision and detail, and the capacity to make and assess intricate arguments, synthesize complex information, and follow detailed procedures and accounts of events and concepts. Students also need to be able to gain knowledge from elaborate diagrams and data that convey information and illustrate scientific concepts. Likewise, writing and presenting information orally are key means for students to assert and defend claims in science, demonstrate what they know about a concept, and convey what they have experienced, imagined, thought, and learned.

Writing, in this sense, takes on the tasks ascribed to it under WAC models (understand, recall, and utilize content-specific knowledge) and the goals of disciplinary literacy (use this information in a context-specific domain, for a specific audience and purpose). And, in thinking about connections to broader literacy goals, Grant, Fisher, and Lapp argue, "[i]n science classrooms the Common Core standards must be coupled with science standards during very purposefully planned, rigorous science instruction" (2015, p. 26).

To layer in an additional way of thinking about where we are at with the integration of writing and science, we share one of the resources that Troy presents in his book *Crafting Digital Writing* (2013). Here, he introduced a heuristic for thinking about writing tasks that involve digital components, MAPS, as outlined in Sidebar 1.4.

When thinking about the MAPS of a writing situation — as compared to being given a list of requirements for an assignment — students are better able to articulate their goals and begin their writing process. Rather than being told exactly what to write, we can engage students in conversations about the audiences they hope to reach, the purposes they hope to fulfill, and the media they plan to employ in doing so.

Considering the many ways, then, that writing can be used in science, we come back to the two overarching ways that we use writing in our classrooms, each with accompanying techniques: writing-to-learn (or writing-to-engage) and writing in the discipline (scientific reports and presentations). We now look at these as we are guided by the inquiry-driven focus of the NGSS and our understanding of the MAPS heuristic, as shown in Table 1.1.

Sidebar 1.4 The MAPS Heuristic for Defining a Writing Task

- Mode — with mode, we think about the genre or general characteristics of the writing, such as how it might be organized, how textual features may (or may not) come into play, and the level of formality one might expect in reading this particular kind of writing.
- Media — with media, as a close companion to mode, we think about the affordances of various writing tools and spaces, including text, image, audio, or video. Media can include word-processed documents, websites, podcasts, infographics, and more.
- Audience — for many assignments, let's face it: the teacher is the only audience. Instead, we want to think about how we can help students move beyond only writing for us as teachers and, instead, think about various audiences beyond our classrooms.
- Purpose — and, again, for many assignments, the purpose is to just get it done and get a grade. But, we want students to think about how they can engage, inform, and argue with many audiences including their classmates and other citizen scientists.
- Situation — with this category, we are thinking both about the situation the writers themselves face (what do they prefer to write, in what formats, etc.) as well as the writing task itself (what do students need to know about the mode and media expected of them, what is the timeline, etc.).

Table 1.1 The MAPS of Writing in Science

	Mode/Media	Audience/Purpose	Situation/Context
Writing-to-learn or writing-to-engage	Primary focus on short, in-class assignments used to develop or deepen knowledge about a specific instructional concept; if additional development of the writing was warranted, it would typically be presented in a traditional, academic essay format and could be extended into a more formal piece.	Primarily for the teacher, though often to be shared with peers is a way to discuss and deepen knowledge; primarily for formative assessment purposes with little intent or expectation to share beyond the classroom.	We can use writing-to-learn activities throughout the class period (beginning, middle, and/or end) to get students thinking about a particular idea, to ask questions, to reflect on what they have learned, or to summarize materials from a presentation, video, or discussion.
Disciplinary literacy	Advocates for disciplinary literacy suggests that students compose pieces attuned to the mode (genre) and media prevalent in the discipline itself. For scientists, this would include composing lab reports, synthesis of information (i.e. review papers, summaries of papers), articles formatted as scientific publications, and graphs representing data.	With the disciplinary focus, the audience becomes one of other experts. The purpose — more than just to process and present information is a way for understanding — is to create an academic argument in a specific discipline. This kind of writing is more formal, and it can be shared with an audience beyond the classroom and formally assessed.	Writing that moves into a deeper, disciplinary focus usually takes more time and involves the use of mentor texts. In order for students to fully understand how to create a lab report or other discipline-specific texts, we need to look at examples. Also, they need to return to their science notebooks and revise the writing that was captured there as first draft thinking.

To put a finer point on it, the new ways in which we are interpreting inquiry-based learning in an era of CCSS and NGSS puts a much greater emphasis on evidence and reasoning than what we had considered before. While it is, of course, reasonable to expect that we can still use writing-to-learn strategies to support our students as they engage with scientific inquiry, it is no longer enough to assume that this kind of writing will be sufficient for them to be successful in either basic STEM education measurements and especially in STEM-related disciplines. Thus, in this final part of the chapter, we outline what we consider are the four key shifts driving current conversations about teaching science today, especially as we integrate writing more intentionally into the process.

Four Key Shifts Driving the Science/Writing Conversation

The experience that Jeremy shared at the beginning of this chapter is, unfortunately, somewhat common. Over many years and across dozens, if not hundreds, of conversations with educators in both K-12 and university settings, the three of us have experienced similar types of frustration, incredulity, and outright resistance to the idea of writing in science. However, the broad trends in literacy education — as well as research on how scientists work as writers — continue to suggest that we will be looking at even more opportunities in the coming years related to writing in science.

In this segment of the chapter, we outline four key shifts that the three of us have noticed in terms of science and writing, especially in an age of NGSS. These four shifts have guided our thinking about every lesson, sample of student work, and discussion of assessment we share in the rest of the book. Put another way, we have seen the following changes in the past decade, all of which are helping us to rethink the ways that science and writing combine in our classrooms:

- First, science and literacy are inseparable, and we are finally coming to believe that the interwoven nature of these two subjects which have long been separated in K-12 and post-secondary education.
- Second, writing and science share a process-oriented stance, and we need to embrace that commonality even more as we integrate the two subjects into our instruction.
- Third, when done well, writing and science also share an inquiry-based stance that, again, we are beginning to recognize and celebrate more and more.
- Finally, digital literacies and newer technologies continue to play a role in science and literacy, and teachers need to pay even closer attention to this than we have in the past.

Here are some extended thoughts on each of these key shifts.

1. Literacy and Science *Do* Mix

The NGSS dedicates an entire section (Appendix M) to illustrate specifically how the NGSS Science and Engineering Practices align with CCSS, especially the Reading and Writing Anchors. In particular, Reading Anchor 7 tackles how an argument should be using diverse formats, including words and visual aids, while the NGSS Science and Engineering Practices emphasize the use of visual representation of data, both when defining a problem and when analyzing results and information. In writing, we can see that the CCSS has a specific set of standards for writing in "Science & Technical Subjects," and a few of the highlights from the 6–8 standards include the ideas that students should be able to comprehend texts as well as, in writing:

- "[P}rovide an accurate summary of the text" (CCSS.ELA-LITERACY.RST.6–8.2);
- "Integrate quantitative or technical information expressed in words in a text with a version of that information expressed visually" (CCSS.ELA-LITERACY.RST.6–8.7); and
- "Compare and contrast the information gained from experiments, simulations, video, or multimedia sources" (CCSS.ELA-LITERACY.RST.6–8.9)

Writing in science is, pun intended, *core* to the expectations in the CCSS. We need to fully acknowledge and appreciate these connections, and take advantage of them as we work with our own students and other educators in building new possibilities for reading and writing in the science classroom.

2. Writing and Science are Both Process-Oriented

Sometimes we look at science textbooks and think that the information there is static, solid, and unchanging, as Wiline argued above. We fail to remember that the knowledge presented in a science textbook had to be discovered in the first place, and that the discovery process is messy (even if it can be somewhat systematic). Given this heavy emphasis on writing in the CCSS, it is no surprise that the NGSS standards, too, have a large emphasis on Science and Engineering Practices, where the process of acquiring scientific information becomes as important as the knowledge itself (the Disciplinary Core Ideas).

This is well represented in the "Three Dimensional Learning" model, where the Core Ideas are just one of three dimensions, along with Practices and Crosscutting Concepts. Really, when reading the NGSS, what strikes us is the de-emphasizing of the minutiae of studying scientific concepts, with value instead placed on how a student acquires this knowledge through the process of the scientific reasoning. This invites teachers to, in turn, encourage their students to make connections between concepts, and to use writing as a tool for inquiry, exploration, and, in the end, explanation. Less time spent on learning meaningless facts in a vacuum means that we can, instead, help students understand the bigger ideas about how systems — from our bodies, to the water cycle, to the working of the entire universe — all work.

3. Inquiry Matters

Both hands-on learning via inquiry as well as modeling scientific processes and interpretation of data are emphasized in the NGSS. Each can be used as a means to acquire knowledge. While it might not be what we may have experienced when we went through school, this new approach to teaching and learning science makes a good deal of sense. If we value the process of acquiring information, then we have to move away from a teacher-driven lesson plan where we simply lecture our students. Yes, students may then be able to fill out a worksheet on the topic, and may even be able to parrot back some key ideas of vocabulary. However, we know that there was no growth for them as scientists, nor any connections made between concepts.

Inquiry-based learning, by contrast, helps them make these connections. However, when students drive the lesson plan by attempting to answer their own questions, the entire process can seem like a daunting endeavor. We are not going to lie: it is. Classrooms — in the midst of an inquiry-driven lesson — are loud, seemingly chaotic places. And, as teachers, we know that it's really tough for us to give up control. None of us want to look unprepared as instructors or ignorant in terms of our scientific knowledge.

Still, we have to engage in the messiness of inquiry.

As Wendy Ward Hoffer argues in her book, *Science as Thinking*,

> [a]mid all the pressures for coverage, inquiry *can* feel like one more thing. But, the good news is, it need not be. By tweaking the work we are already doing in our classes, science thinking skills can walk hand in hand with content learning.
>
> (2009, p. 25, emphasis in original)

Inquiry-based instruction can be as simple as starting a unit with a question, and letting students use various media and design experiments to answer that question. Yes, it is slow and deliberate; students need time to get to the take-home message on their own, to be gently redirected by the teacher as needed. But the end result is theirs, unlikely to be forgotten, with empowering skills gained along the way.

4. Technology, Writing, and Science are Intertwined

If you are a teacher that has come to this book having already read either *Create, Compose, Connect* or *From Texting to Teaching*, you know that Jeremy and Troy have long been interested in the ways that technology in writing intersect. Science instruction, then, offers the perfect petri dish in which to explore even more connections between writing and technology. In particular, you will see examples throughout the book where we are working diligently to incorporate the International Society for Technology in Education (ISTE) Standards for Students in thoughtful, productive ways.

In particular, when the most recent version of the student standards came out in 2016, ISTE made a point to discuss how they were shifting the focus from "learning to use technology" in the earliest iteration (1998) and "using technology to learn" in the previous iteration (2007) to a more purposeful, "transformative learning with technology" in the current version. We are trying to do the same. As you review the strategies that we share throughout the book, our hope is that you will notice how students are using their mobile devices and school-issued laptops to be more purposeful with their data collection and analysis as well as with opportunities for WTL, and to share discipline-specific pieces of scientific writing. We try to mention the standards, as appropriate, when there are technology connections to be made.

Conclusion: More Alike than Different

In summary, we have been fortunate to have done quite a bit of thinking together about writing, both in the history of and process of doing science, as well as about writing across the curriculum and writing-to-learn. All of this has prepared us for a deeper dive into discussion of our own teaching practices, which we will explore throughout the rest of the book. We know that science and writing are tied together in nuanced, complex ways, and helping our students engage in inquiry-based learning will be the best way to help them grow as writers and as scientists. In the next chapter, we look at the process of integrating science notebooks into both middle school and university science classes.

2
Science Notebooks

While we are not 100% certain when the first notebook was used, we feel pretty comfortable making the case that writing in a notebook has been part of *doing science* for hundreds of years. Asking and exploring, through the act of writing, are part and parcel of what scientists do.

Many examples, including the copious numbers of notebooks from Charles Darwin, Marie Curie, Thomas Edison, and Leonardo da Vinci, all demonstrate that scientists spend a good deal of time writing, and this writing can range from just a few words to full paragraphs, essays, and more. As we have noted already, Jessica Fries-Gaither has collected examples of about a dozen scientists' notebook pages from publicly available images, complementing those entries in prose in her award-winning book for children, *Notable Notebooks*. As she demonstrates, scientists of all eras, from all disciplines, use notebooks to write their ideas, in process, and then move those ideas into more formal writing.

The notebooks also contain, not surprisingly, many drawings; from initial sketches to generalized models to detailed drawings, all sorts of images are found within their pages. We have formulated some of our thinking about notebooks from *Composing Science: A Facilitator's Guide to Writing in the Science Classroom* (Elliott, Jaxon, & Salter, 2016). In their book, Elliott, Jaxon, and Salter describe science notebooks as spaces where "students quickly adopt a scientific approach" where they can "make inferences, show evidence of metacognition" (17) and "utilize diagrams" (18). If scientists have learned anything, it is that writing is flexible, and what starts in our notebooks can lead to other spaces. This is the power of writing, and drawing, in science.

A quick note to begin. We choose to describe this kind of writing as "science notebooks," even though we can understand that others might use "science journals" to describe this genre. We intentionally use "notebook" — in both the middle school and college classroom — to describe the *tool* (the item itself) and the *activity* (to notebook, or to write). We might just as easily say "It's time to notebook" as we would say "It's time to get out your notebook." In other words, students hear "notebook," and they know it is time to write, draw, and think. And, they know that unlike a typical journal prompt that they might be given, they need to do the work of expressing their thinking through words and images.

Also — and perhaps more importantly — in any web search for "science journals," you are likely to find dozens of worksheet-style options, which are exactly the opposite of what we are looking for in a notebook that students develop on their own. "Journaling," sometimes, has a negative connotation for students who do not see themselves as writers and, while we have used the term "journals" to describe these practices in our teaching in the past, we have, in recent years, shifted to use "notebooks," and we will continue to do this in the future.

In this chapter, we invite students to use their notebooks as a space for exploration and inquiry, for making observations and asking questions. We introduce the kinds of informal writing that can lead to more significant writing in science. In our classrooms, this informal writing is important and useful, though it is not the kind of polished work that would normally appear in a final report on their work. As we will see in other subsequent strategy chapters, we will outline what the strategy is, why we use the strategy, specific teaching moves, some considerations for assessment, and next steps for getting started with the strategy in your classroom.

What Science Notebooks Are

While everyone is entitled to use — and define — science notebooks in whatever way makes the most sense in their own classrooms, here, we borrow the definition from *Composing Science* (Elliot, Jaxon and Salter), and extend it slightly: "Notebooks are a place where students have the freedom to try on ideas without being judged if those ideas are wrong" (p. 28). This means that students can use their science notebooks for both formal and informal writing, choose the style that works best for them, and include visuals such as pictures and diagrams, which are critical elements for the inquiry process. Elliot, Jaxon, and Salter specifically mention writing-to-learn as an opportunity and outcome of using notebooks in science, with which we agree.

We can say that Elliot, Jaxon, and Salter would define notebooks as a flexible, student-centered space for informal tasks that include writing-to-learn activities like quick writes, summaries, questions, and observational notes. Building on from this definition, the three of us agree that notebooks play a critical role in student's development of ideas. So, we do not have much to add to this definition, though there are a number of specific activities that we might suggest for our students with their notebooks. Sometimes, these are intentionally scripted moments for writing that we drop into our lesson plans. Other times, the opportunities for writing just seem to present themselves as we teach. For instance, we might ask students to do the following:

- Notetaking on class discussions, readings, and videos;
- Making observations about the world in a "natural history" style of writing;
- Posing inquiry questions or rethinking previously written questions;
- Summarizing points of study for future quizzes and tests;
- Documenting data from scientific investigation, especially experiments; and
- Comparing and contrasting their own data with the entire class's data.

Additionally, we suggest that students can use their science notebooks as a place to take lecture notes, and we will elaborate on strategies for accomplishing that task in a meaningful manner. There are many ways that we use notebooks and notebooking, sometimes even moving between two or more ways in one class period. In this sense, we generally see the MAPS context for science notebooking as outlined in Table 2.1.

Table 2.1 The MAPS for Writing with Science Notebooks

Mode	Brief, in-class "writing-to-learn" or "writing-to-engage" (as short as 1–5 minutes, or even up to 10–15 minutes in the college classroom).
Media	Most often, we encourage students to use a pencil and their physical notebook, although we do not discourage the use of pens, as we will elaborate below.
Audience	For the student him/herself, primarily, as well as for the teacher as a formative assessment, secondarily.
Purpose	Usually at the lower to middle levels of Bloom's taxonomy (remembering, understanding, applying, and analyzing), though sometimes we do encourage students to write extended pieces that push toward higher levels of Bloom's taxonomy (evaluating and creating) and, ultimately, to make connections between concepts we are learning.
Situation	As noted in the "Mode" above, the situation for notebook writing can vary, happening at various points during class time. And, as much as we would wish for students to take the initiative to constantly use their notebooks, we would be disingenuous if we didn't say that we are, often, the ones who must prompt them to do so.

One of the reasons we dedicate an entire chapter to notebooking is that notebooks provide an approachable tool to introduce systematic writing in the science classroom. Notebooks are relatively easy to implement at a low cost. Students don't need much more than a plain old notebook and a pencil to start notebooking. In recent years, Wiline started using an artist sketchbook, large format and white blank pages, as science notebooks. She likes the freedom of the big open pages without lines, and finds that it can inspire students to sketch in addition to writing. Some science teachers might be reluctant to adopt notebooks, usually because of the worries linked to grading, but we provide some tips to address this concern at the end of this chapter. Plus, there is a rubric Jeremy has included that he uses in his own classroom to help anyone who may need something for their own classroom.

In the remainder of this chapter — as we will do in subsequent strategy chapters — we look at how this strategy plays out in both Jeremy's middle school and Wiline's university classrooms.

Why Use Science Notebooks

Jeremy's Perspective

My students use their science notebooks for inquiry, experiments, notetaking, drawing, and as a place of reference for future quizzes and tests. For that reason, at the beginning of the year I recommend to students that they purchase a 1 ½ inch three-ring binder and a sketchbook. By using a sketchbook, the students are more likely to draw pictures, charts, and tables along with their writing. In the past, I have used just regular lined paper, but prefer the sketchbooks. Students use the binders to store handouts, quizzes, and tests.

I did not find my groove for science notebooks until I read *Composing Science*, which has been mentioned above. I learned that students need to take ownership of the work they put in their notebooks. As teachers, we need to give them the freedom to compose any way that helps them learn and that means allowing them to make lists, use bullet points, or draw pictures. Furthermore, it became clear to me that students didn't need to just use notebooks to take notes in class or to be used as warm-ups. Notebooks should be continuously used throughout class with experiments, field studies, and in class projects. *Composing Science* opened up the world of science notebooks to both me and my students.

Before I dive into what I do in science, I would like to give a sense of where I have been with notebooks in my English class. One of the first things I learned as an English teacher was to implement some sort of writer's notebook into my English class. I was taught early on in my teaching career that this allowed for students to write every day. Over the years, my practice of using notebooks in my English class has evolved substantially. I started with making students do the prompt every day and then I would grade them at the end of every week. This was, in one word, exhausting.

Then, I transitioned into allowing my students to be more open in their notebooks with their language and writing about more personal aspects in their life. This idea came from reading *Freedom Writers* (Gruwell & The Freedom Writers, 1999) and watching the movie starring Hilary Swank. I also made sure that students understood I had to report certain things I might read in their notebooks. Furthermore, I backed off my grading from every week to, instead, only twice in a marking period, and that was much more manageable. Still, they were more about freewriting to build fluency and stamina, and were much more about personal writing than anything that would really demonstrate a students' thinking.

When it comes to science, I had to relearn everything about what notebooks were and how to make them useful for my students. On the one hand, notebooks are a place for the kinds of writing that we would expect scientists to do, with many data points described, scientific terms defined, and drawings that support what they are observing. In this sense, as teachers, we often feel as if we must have students push aside writing about their feelings in their notebooks. In short, we want students to write-to-learn when they use their science notebooks and, while they can still express ideas and their opinions, feelings aren't necessarily important when it comes to exploring and making scientific observations.

On the other hand, the more that the three of us have worked together in professional development and in writing this book, the more we became aware that letting students use the notebook as a place to express thoughts and feelings matters, too. I tend to lean away from this though. It doesn't mean I take the freedom away from them to write in a way that best expresses themselves; however, they are more likely to compose something akin to a personal narrative in their Language Arts class, but not in science. In a science classroom setting, students are inquiring about information they have gathered and will be making claims, providing reasons for their claims. Notebook writing in science encompasses more informational and argumentative qualities of writing, and is not simply about feelings. Yet, the wonder of inquiry remains an important aspect, and if students do occasionally write about their emotions, that's alright with me.

Developing my own ideas about using science notebooks has been a journey. In Sidebar 2.1, I mention five particular things I have learned about implementing science notebooks into my classroom. This past year was the first time in my science class where I have gone away from using notebooks as just a warm-up and, instead, have allowed students to engage in open

Sidebar 2.1 Five Things Jeremy Has Learned About Having Students Use Science Notebooks

1. Students learn that science notebooks are a space they can express their ideas and not be judged. Because of this, they are more likely to share in class.
2. Notebooks lead to more inquiry questions from students when they are used for more than just answering questions or taking notes.
3. Grading science notebooks really is not that difficult if you design a rubric that best fits the needs of the students.
4. Science notebooks can accommodate those students who may do a better job of communicating what they have learned through pictures.
5. Science notebooks alone can help lead to better literacy skills in students.

inquiry. When it comes to notebooks, in both science and English, students are going to have their own style of how they compose their thoughts, how they put words on the page, and also bring in illustrations as well. From just writing numbers in columns to drawing intricate diagrams, scientists have a style, too. As teachers, if we nourish and allow our students to create their own style (within the flexible structure we provide), students can excel and become better writers across the curriculum. We'll explore some examples of my students' work below, but first will turn to Wiline to hear about her experiences with notebooks.

Wiline's Perspective

As we have tried to make clear, writing is an essential part of a scientist's life, especially when it comes to communicating findings with the rest of the community. Prominent scientists are often those that can write in a compelling fashion, having mastered the art of explaining complicated processes in a way that most folks can understand.

As just one example, I am amazed by the work of Robert Sapolsky, who manages to present intricate relationships between hormones and human behaviors in works such as *Behave: The Biology of Humans at our Best and Worst* (2017). Yet, the training that we receive in scientific writing can be very sparse. A case in point is that I did not have to take any writing classes as part of my education until my doctorate program and, even then, it was just one course. Compare this to the dozens of science courses I had to complete. Science writing is often self-taught (and painfully so, at that) in an effort to achieve some measure of professional success.

To reverse this trend, we as instructors now tend to embed more writing within undergraduate science courses, and science notebooks become an essential tool. As any skill, writing requires practice. Science notebooks provide a safe space for such practice, especially if students know that the writing is not directly graded. Notebooks provide an opportunity to write frequently, to go through the process of picking what to write, how to write it, and to experience the painstaking work of moving from thought to words, and from words to sentences.

In addition, notebooks do not have to be simply about writing: students go through the process of deciding when it is actually better for them to sketch instead of write; that is, drawings are both invited and encouraged. Over time, notebooks become inherently personal: no two look alike, and giving students such freedom is essential in generating meaningful products, as we will show later in this chapter.

From a pedagogical perspective, science notebooks have an added benefit of providing an opportunity for students to grow, both as writers and as scientists, and to witness such growth. One of my favorite activities, toward the end of a unit, is to make students go back and read their very first notebook entries. Students are often surprised by how much they learned. In addition, they realize what specific entries have been more useful than others, all without a need for a grade or an external evaluation. Notebooks provide a track record of thoughts for an entire string of concepts. Looking back at old entries allows students to draw connections between concepts, which can be such a difficult thing to accomplish in a classroom. Lastly, notebooks can generate a sense of long-lasting pride that can be sometimes missing in a science classroom — and an overall school experience — that is mostly full of short-term experiments and assessments. In Sidebar 2.2, I explore some of the things I have learned about using science notebooks in my classroom.

Sidebar 2.2 Five Things Wiline Has Learned About Having Students Use Science Notebooks

1. What students enjoy in my science classroom, lecture, or field activity is not always what I think they will enjoy! My perception can be off, which I can realize by reading reflection pieces in my students' notebooks.
2. Students vary drastically in their notetaking abilities and skills. I am always surprised, when I collect notebooks and look at lecture notes, at the differences in note quality between students. Some students have amazing lecture notes, with all diagrams redrawn, a logical framework clearly apparent, colors to help them navigate content. Other students have barely only notes for a given lecture, besides maybe a few vocabulary terms. Yet, I have learned that I cannot use this difference as a good predictor of exam success: some poor notetakers still manage to do very well in difficult exam, which reminds me that we teach students with very different learning styles.

3. Reflection pieces and opinion-based entries are my favorite sections to read in my students' notebooks. It's fascinating to be allowed to learn a little bit more about my students in such a personal manner. However, I have been told from students that these more open-ended entries that were reflective were often their least favorite entries to write! I wish my students knew how fun it is to read some of their writing, and how much my teaching benefits from getting "raw" feedback.
4. Repeating a specific activity or notebook usage many times within a unit/semester is critical for students to develop their own "groove" in their science notebooks. It can take students a little bit of time (4–5 entries of one type) to start to really gain from the exercise. This becomes pretty obvious when reviewing notebooks.
5. Sometimes, what I think is a clear writing prompt is not! I have sometimes provided what I thought was a clear writing prompt, and quickly realized when looking back at notebooks that all students interpreted the prompt in a very different way. That is not necessary a bad thing and can generate some interesting entries and conversations.

How to Teach with Science Notebooks

Jeremy's Perspective: A Case Study with Photosynthesis

From the first day of class, my students engage in observations and experiments. As we've discussed already, another identifiable similarity between science and literacy is when students are carrying out an experiment and are fully engaged in the process. For example, in our first unit in 7th grade, students grew radish plants. They had to design an experiment where they had a control plant and an experimental, manipulated plant, both tracked over time for growth.

Students were given a list of materials available to them, and were left to decide as teams of two as to what they would change between the control and manipulated plants. The idea behind the experiment was for students to discover how a seed actually grows into a plant and manipulate the variables that might contribute to the process, connecting directly to the NGSS standards on Growth, Development, and Reproduction of Organisms (MS-LS1-4 and 5, MSLS3-1 and 2, and MS-LS4-5).

To offer an example, one group of students used Powerade instead of water when it came to their second plant. That particular group wanted to see if the plant would grow as much, or less, with the use of a sports drink instead of water. I had to give them credit, because at least they were thinking about the possibilities of electrolytes. Other groups explored over-watering and under-watering, another used lemon juice to see if it might help and, yes, one group even tried a Tide pod. What can we say, they are middle schoolers after all!

As you can see in Figure 2.1, Braleigh and her group wrote a notebook entry documenting what they were going to do throughout the experiment. Starting on September 11, 2018, she documented her process of creating a controlled experiment where the "controlled" seed is described in the left column, while the "experimental" or, in the students' word, "manipulated," seed, is described in the right column. Then, on September 12, she makes some predictions about what she expected to see happen based on the differences between the two treatments. Finally, on September 17, after nearly a week of growing, Braleigh documents what she has seen.

It is important to note two key elements of this response. First, Braleigh is using scientific terminology that we've discussed over the course of the week (such as "germinated"). Second, she makes an important transition when she states that "A seed is made from matter." And, while we will talk more about assessment of the science notebooks later in this chapter, suffice it to say for now that the student earned 9/10 because she presented her experimental

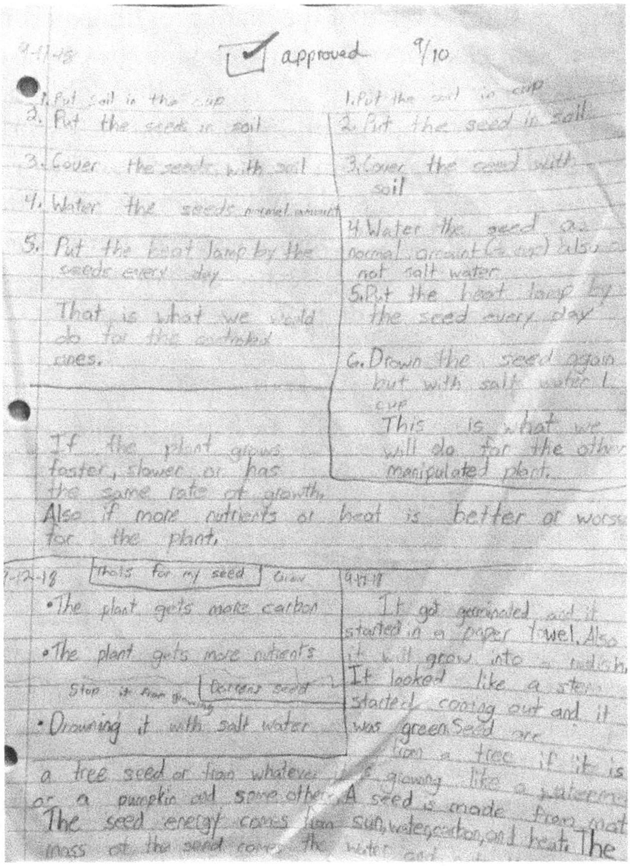

Figure 2.1 Braleigh's Initial Notebook Entry for Seed Experiment

design and had a solid plan for how she was going to do. The missed point comes from not discussing that the data were going to be measured throughout the experiment, specifically the mass of the plant. And, even with the one missing point, she (and her parents) still gave me permission to share this in the book!

When we compare the process of what has happened in this science notebook to what typically happens in an English classroom, there are similarities. Of course, students brainstorm at the beginning of the writing process and create an outline; this helps them organize how their writing is going to look when they start the drafting process. As students proceed through the drafting process in an English class, their writing is constantly evolving. Students edit and revise, and oftentimes they have multiple drafts. From a science perspective, students go through a similar process while conducting and documenting their scientific study. Often, students begin conducting their experiment and then — as they are collecting data —realize that they need to revise their experimental design. This could mean that they need to change their variables, or that they need to change the type of data they need to collect.

In the example outlined above of the plant growth experiment, students were using their notebooks not only for notetaking, but to also record their data during the experiment. Figure 2.2 shows Braleigh's data collection throughout the time of the experiment. In particular, it is worth noting that she set up their data collection system like a calendar. It's worth noting, too, that Braleigh and her partner were the only students to do that, a strategy that other students realized they should have been doing when we got further along in our work. Finally, note that she "grayed out" the spaces for data collection, as their own plants had died. Still, other comments were added (as shown in the margins).

Besides the examples above, other kinds of writings and drawings that I might see in my students' notebooks include very detailed drawings of specific objects of study. In Figure 2.3, Braleigh drew a picture of a leaf that she gathered on a mini-field trip on our school property. At the time, students were trying to discover where the stomata on a leaf existed; stomata (the plural for stoma) are very small holes located in the back of leaves for the gas exchange, necessary for both photosynthesis and respiration. After gathering the leaves they needed, students returned to the classroom and used microscopes to observe where the stomata were located. Braleigh not only drew the leaf, but made a detailed drawing of what she found under the microscope. She included the cell wall, chloroplasts, and what she thought could be identified as the stomata. What I particularly found enlightening is that she included a detailed description of her findings.

Figure 2.2 Braleigh's On-Going Notebook Entry for Seed Experiment

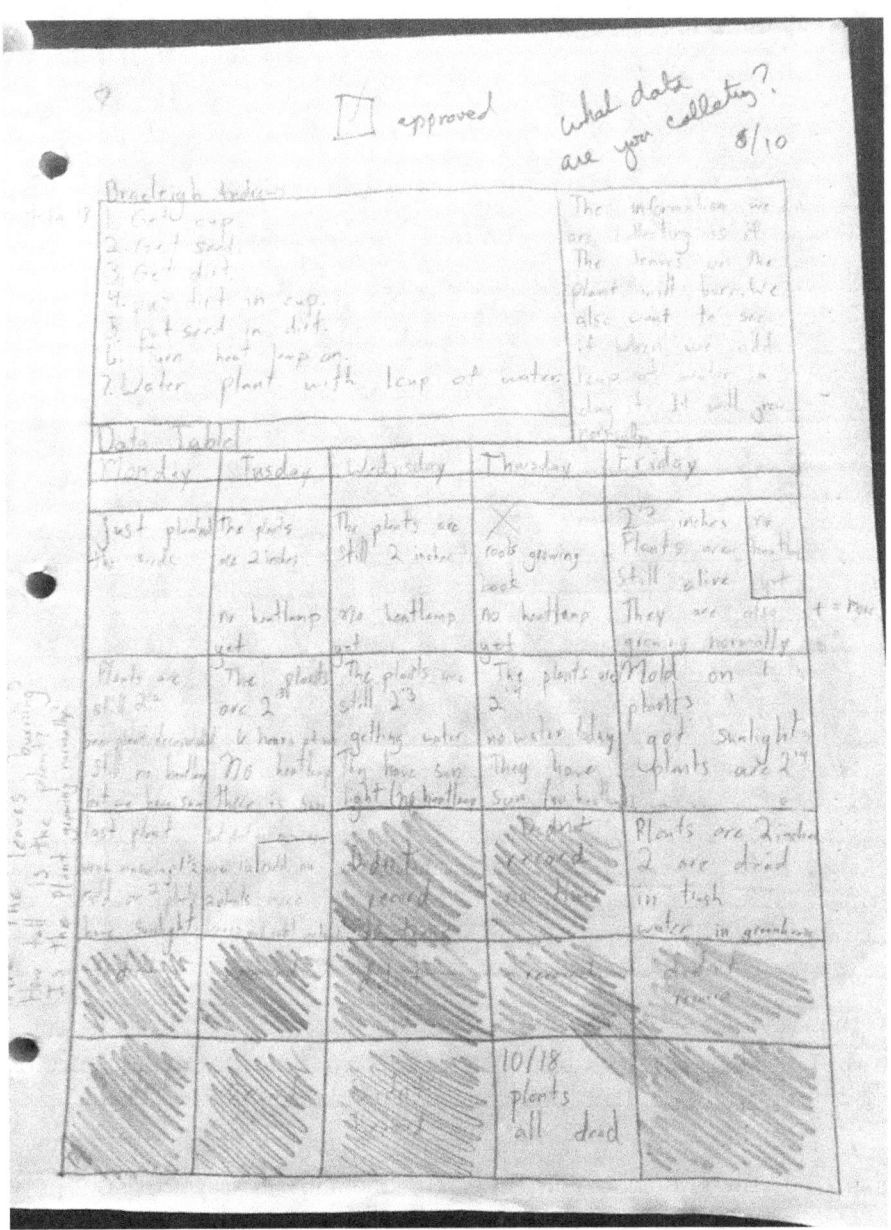

Notebooks have become an integral part of my science classroom. It is no longer something that I only use as a five-minute warm-up at the beginning of my class, like I have done in the past. Science notebooks have developed to a level that students can feel comfortable using them for anything in science class, and my young scientists literally carry their notebooks everywhere

Figure 2.3 Braleigh's Notebook Entry for Leaf Observation

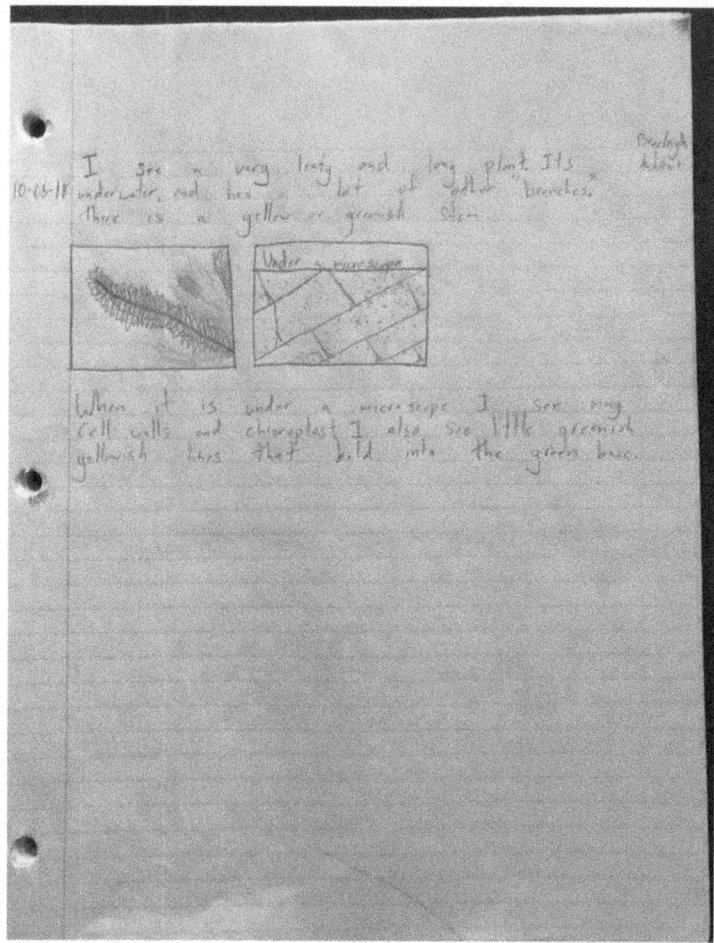

they go in science. It has become one of my most important tools for science. Below, in Sidebar 2.3, are different ways I have incorporated science notebooks throughout the plant/photosynthesis unit.

Sidebar 2.3 Jeremy's Notebook Entry Ideas Throughout the Photosynthesis Unit

Science Notebook Entries
1. Create an experimental design for your control plant and your variable plant.
2. Draw a picture of your experimental design and label it.
3. Outline your steps and procedures for step-by-step so that other scientists could replicate your experiment.
4. Reflect on your experimental process. What worked? What did not work? Explain what you would change.
5. Create a table and define the terms *matter* and *energy* in one column. In a second column list examples of matter and energy.

Wiline's Perspective

When I think about how I incorporate science notebooks in my classroom, there are four general usages that I keep coming back to:

1. to make observations about the natural world;
2. to record data;
3. to brainstorm about experimental design, including expected results; and
4. to record notes on readings, lectures, or videos.

These uses are not mutually exclusive, but each does serve different learning goals. As such, all uses do not, in fact cannot, happen at once, especially if students are new at notebooking.

Below, I break the practice of science notebooks into the categories that I use any time I implement a pedagogical tool in my classroom. With each use, I am considering:

- when I implement this tool throughout the semester or academic year;
- where (in the classroom or outside class time);
- how long it takes to implement;
- what kind of directives I need to provide to my students;
- what my goals are; and
- what are some potential follow-up activities I may be able to do.

In the sections below, I explore four general categories of notebook entries (observations, data, brainstorming, notetaking), and provide annotated examples for each. While these examples are from older students, our companion website provides some additional examples for early elementary grades <https://jeremyhyler40.com/science-and-literacy/chapter-2/>.

Observational Entries

This type of entry relies on observations about the natural world, in an open-ended format, often happening at the beginning of new units. We might walk outside, even if just for 5 to 10 minutes, to capture initial ideas. Then, students are invited to write a more complete entry at home. To focus their attention, I typically provide a prompt that invites students to think about one specific concept; I might suggest that they "choose a flowering plant that you find interesting, and make observations." Students have the option to draw or write, and I usually encourage them to do both. It is interesting to have students use ink instead of pencil for such type of entries: with ink, students cannot erase their entries to modify them simply to please me as a teacher, and entries become more personal and spontaneous, with the permanence of ink actually setting their creativity free.

This exercise increases students' power of observation, regardless of how familiar they may be with aspects of the natural world. Moreover, it develops the process of inquiry, forcing them to slow down and appreciate details about specific organisms, in this case plants. I've tried this activity with students as young as kindergarten and, of course, with my college undergraduates, and all have found value in it. As a follow-up activity, I may ask students to then begin posing questions about what they have drawn and written, moving into inquiry questions and hypotheses. We also discuss why they may have chosen one type of organism over another, and then compare and contrast what we have discovered.

The following images show different notebook entries from students, all using notebooking as a tool to increase observations. In each, you will see different types of details; none of them are perfect, but each brought new insights to the learner. van Dijk, Gijlers, and Weinberger (2014) did a study with elementary students and discovered that, when used with a protocol to scaffold their learning, the drawing process can help students deepen their scientific knowledge.

The first example (Figure 2.4) comes from the classroom of Erin Wright, a 5th-grade teacher, whose students were looking at plant succession along a

Figure 2.4 Invasive Species in a Michigan Woodland

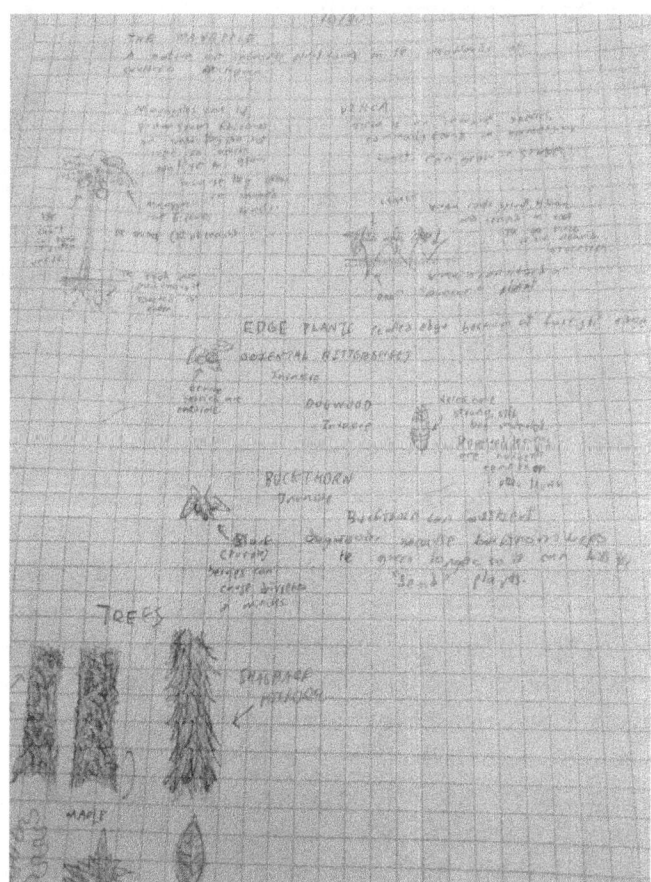

woodlot near their school. In this case, her student Ivo Saran notes many features of the mayapple, vinca, oriental bittersweet, dogwood, and buckthorn, all invasive species that were encroaching on the natural habitat occupied by the oak, maple, and shagbark hickory trees. The students were guided by docents of the woodlot, so they had ample opportunity to take extensive notes and ask questions, even delving into the root structure and leaf physiology. Ivo's entry shows extensive detail and demonstrates a complex understanding of the role of invasive species in the woodlot, along with drawings to help identify species later.

I've pulled one of my own notebook entries as an example of what can be done at the undergraduate level. I was an avid notebooker, and I believe to this day that notebooking for my Easter African botany class is the only way I managed to pass the course! In Figure 2.5, you can see that while the

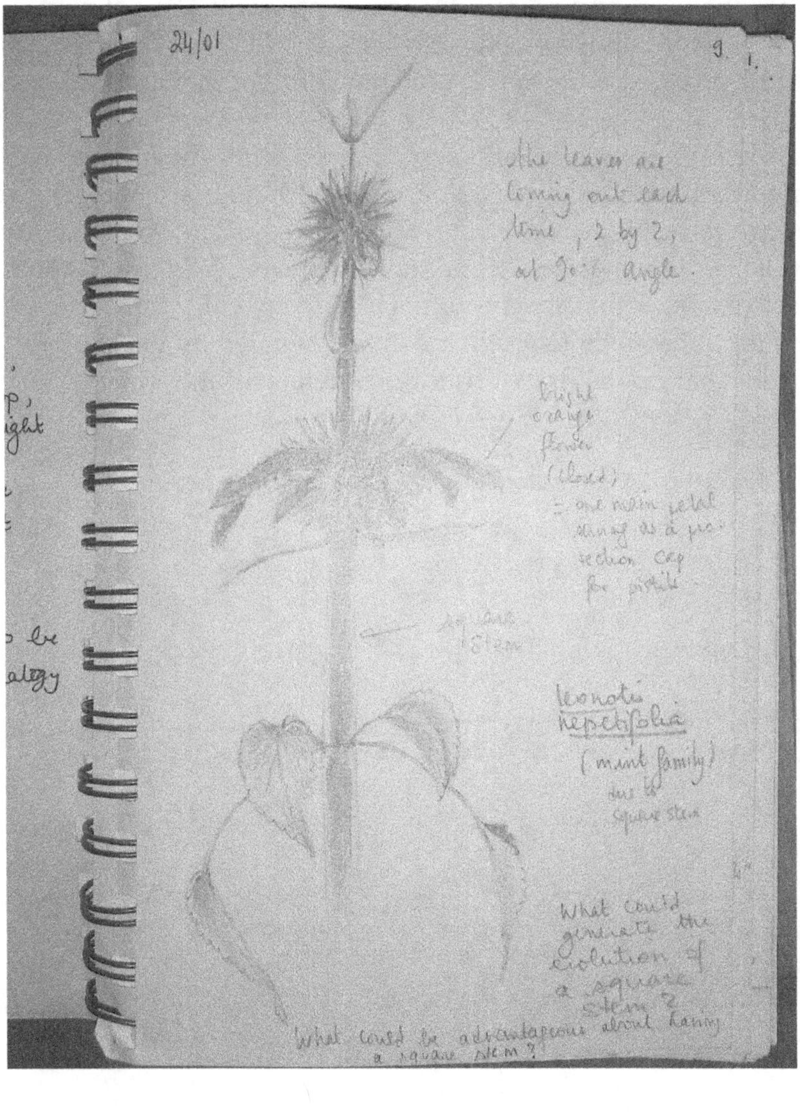

Figure 2.5 Wiline's Undergraduate Notebook Entry on an East African Plant

drawing is interesting, most of the key aspects of the entry are present in the notes next to the drawing, depicting leaf placement, color, stem shape, and scientific name. I even posed a question: "Why are the stems square?" Most importantly, the process matters here: spending the time it took for me to draw this meant that I could now recognize this species on any test, be in the field or in the lab, as I knew its intimate details.

Data Recording Entries

Another kind of notebook entry that I use with students is to have them document what is happening during experiments. In this case, we are always "on" as scientists in the classroom and students need to be able to document the results of their data collection in order to, later on, analyze them in a meaningful fashion. This can take just a few minutes for a short experiment or up to an hour or even more for longer experiments. Sometimes, data collection occurs over multiple class sessions. In whatever manner it unfolds, some of the prompts in Sidebar 2.4 can be useful for moving students' thinking forward.

Providing directions for students, at least with this kind of writing, is very important. Rather than a full-out prompt though, we need to give them just enough to begin their thinking. Sometimes, I share what I'm expecting them to do by drawing a model on the whiteboard or in my own notebook, using the visualizer for display. I often divide my entry into several categories including the methods used, the data generated by the experiment, the conclusions that can be drawn from such data, and the summary of results

---

Sidebar 2.4 Notebook Prompts for Collecting Data During an Experiment

During an experiment, students need to document what is happening. There are two types of data entries that I use in such cases:

1. Data entries that are general. I might give students prompts such as:
    a. What are you noticing about _____?
    b. How would you describe the process of _____?
    c. What happened at the beginning/middle/end of the experiment?
    d. Did [variable x] have a visible effect on [variable y]?
2. Data entries that keep track specifically of the dependent variables we are measuring. I might give students prompts such as:
    a. Pick one dependent variable, and record how it changes over time of the experiment.
    b. Record what is happening every minute of the experiment.

from other groups. If possible, I let students decide how to organize their data within these categories, and sometimes that leads them to record their data in a way that may, unfortunately, be less than useful later on. This reiterates the point that they must be specific and systematic in their data collection, and so it is a lesson we learn through trial and error early in the semester.

Ultimately, the goal is to have students experience what it means to record data from the experimental procedure so, the next time they run a procedure, they can be even more intentional in the way they designed it and in the way that they set up their data collection. If we were to just give them a lab worksheet, they would not be able to go through this thinking process on their own, and this fundamental aspect of science inquiry would be lost on them.

We can see this process of organizing data for data collection in Figure 2.6. Jessica, a high school student taking an introduction to field biology course, followed her own logical flow on the left side of her science notebooks. She was recording the behaviors of Eastern chipmunks outside while they are feeding at a patch of seeds. She used her blank page to first take a few notes on the introduction to the lab about chipmunks, and then she divided the data collection by type of dependent variables (which she even decided to number "1. Number of heads up," "2. How long they stay," and "3. How far they travel"), which was further divided by individual chipmunks. Below this raw data collection, Jessica started to actually summarize her data in averages.

Moving from the raw data into graphing makes perfect sense at this stage, as seen in Figure 2.6 on the right side of Jessica's notebooks. Jessica created two different scatterplots with the data, depicting the effect of burrow distance (her x-axes) on different dependent variables (time spent at the foraging patch on top, and number of head ups at the bottom). Notice the careful labelling of axes, including the units. And we can reflect here on the quality of the science processing that Jessica had to do to generate these graphs — she had to process the data on the left, decide how to organize it, what to actually graph, and what type of graphs to use. This would have all been lost if, instead, she had been handed out a worksheet with the graph axes already provided.

Sometimes, I invite students to do this individually, sometimes in small groups, depending on the time we have available in class and my overarching goals for the lesson. And, perhaps most important of all, we can invite students to think about the process. Based on what they recorded — or failed to record — during this experiment, they can consider ways in which they might approach the next experimental procedure and data collection opportunity.

Figure 2.6 Jessica's, a High Schooler, Data Recording Entry on Chipmunk Behavior

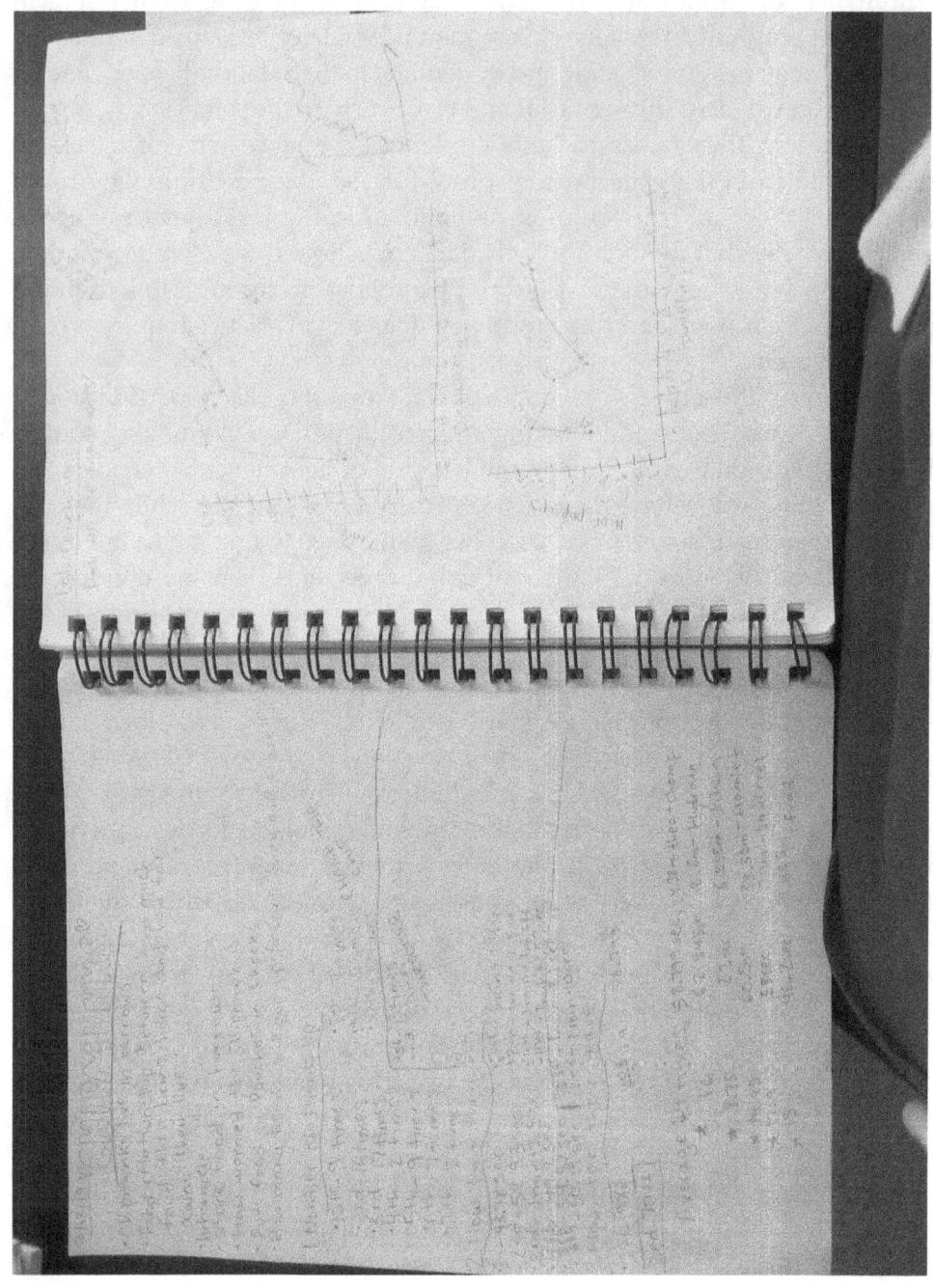

Brainstorming Entries

As simple as it sounds, we often forget that students can use their notebook for brainstorming. This can include brainstorming about results they have discovered through their data, as well as giving them an opportunity to develop new hypotheses and predictions. Moreover, they can reflect on their procedure and think about how best to design the next experiment. This strategy often comes as a complement to the ideas described in the "data recording entries" section above, and most often happens in the classroom. It serves as a nice way to transition from one unit to another, and can be as short as just a few minutes. Or, depending on how much time we have available to devote to the task, it could take up to thirty minutes or more.

In order to do this kind of writing, I typically provide some directions, and I may use prompts like the ones in Sidebar 2.5:

Sidebar 2.5 Prompts to Help Students Begin Brainstorming

- Jot down three different ecological questions about the ecosystem we are about to study.
- What type of questions would you want to ask in this experiment?
- Come up with a few different hypotheses about this experiment.
- What are some specific predictions (at least 3) that may occur in this experiment?
- Describe three strengths of your experimental design (or data collection). Describe three weaknesses.
- If you were to redo your experiment, what would you change?

Reminding students that this is not a formal piece of writing, and that brainstorming is meant to be creative and open-ended, they feel a bit freer (I hope!) in their willingness to write. It allows them to gain practice in fluency with the thoughts before turning them into a more formal report at the end of the unit.

Figure 2.7 displays an example of a high schooler, Molly, who brainstormed for a class project on hemlock trees. Her notes are spontaneous, full of facts and concepts she read about (the left page of her notebook) and questions she was trying to answer (right page). She was learning to handle a lot of new vocabulary specific to plants (such as allopathy, the production of chemicals through roots, which prevents the growth of other plants), and applying that vocabulary to her specific project.

Notetaking Entries

For the use of notebooks to record notes on readings, lectures, and videos, the process is pretty much self-explanatory and the examples a bit mundane. However, having students turn in their notes for assessment can be highly revealing, both for us as instructors, and for the students. I recently started

42 ◆ Science Notebooks

Figure 2.7 Molly's Notebook Entry Brainstorming Ideas about Hemlock Trees

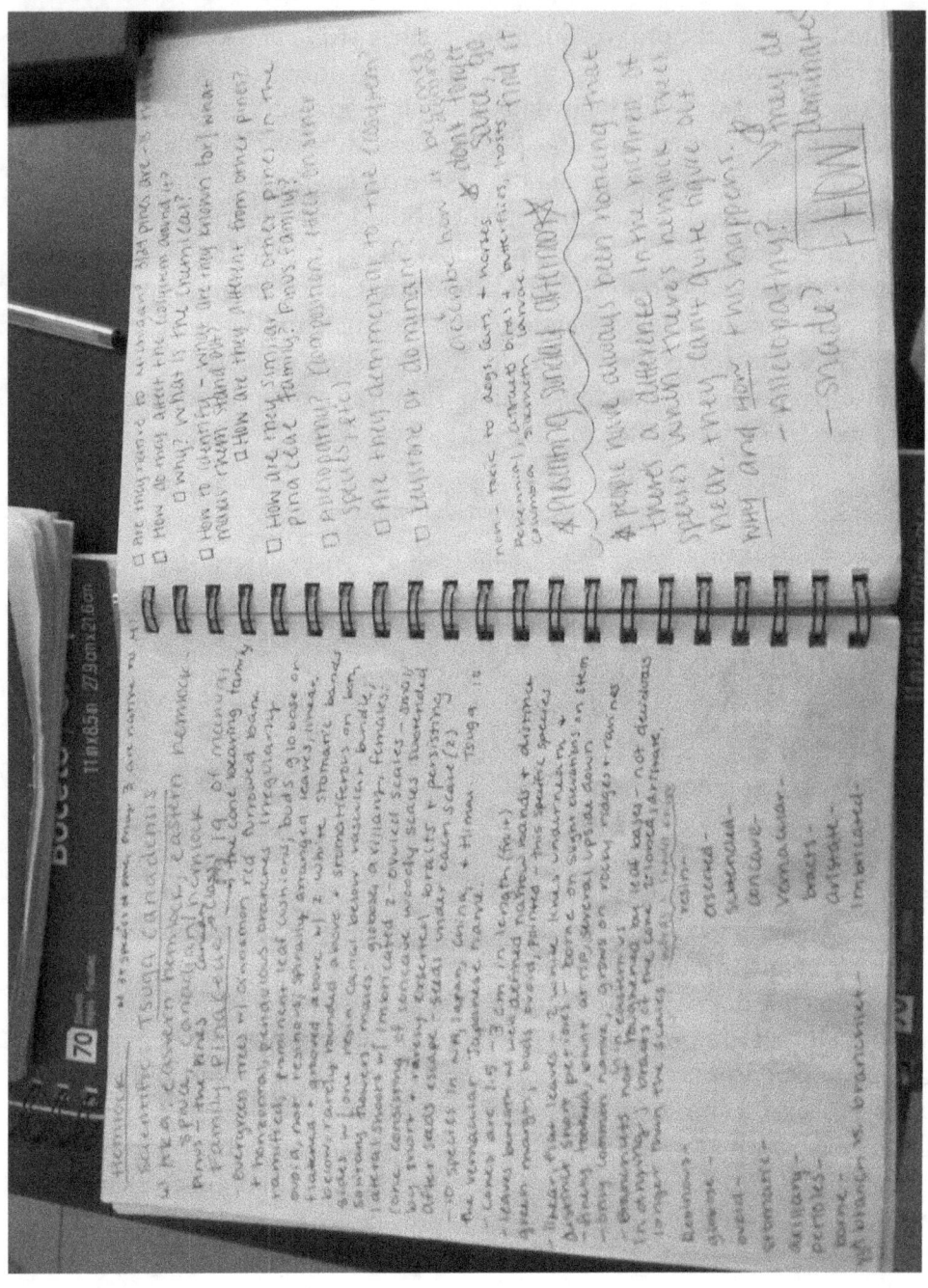

to systematically look at lecture or reading entries in my classes and realized where students were needing more help taking notes, and where notetaking was sufficient. For students, receiving feedback on their notetaking entries was new: many didn't realize they were omitting lecture concepts, or needed to not just write, but also draw any diagrams. It can be powerful for students to witness their growth in their notetaking skills from their early notebook entries to their last notebook entries.

A number of notetaking strategies exist within the broader world of writing-to-learn. In particular, the concept of "Cornell notetaking" has been around for a long time. Students try to document the main ideas from the lecture in one column while, in the other, jot additional questions and ideas. A quick search of "notetaking strategies" yields millions of hits (some, of course, to some site that will sell students notes!), and a few other popular strategies include the classics of outlining and mind mapping, as well as some more recent innovations such as "sketchnoting" (Heick, 2013) and using a "bullet journal" (Nguyen, 2016).

If you do choose to invite your students to move to their computers or tablets to take notes, there are a variety of tools that can help them. Some apps that have traditionally been used for typewritten words (such as Apple's Notes on iOS and Microsoft Word) allow for handwritten notes when used on a mobile device, and the iOS app Notability has long been a favorite. Also, Microsoft's OneNote is becoming better and better at allowing typed and handwritten notes to be combined. Other apps, like Google Keep or Zoho's Notebook can serve as robust spaces for saving and sharing smaller chunks of notes. Finally, if you want to have students move their thinking into some Good Drawing-based graphic organizers, Matt Miller's *Ditch That Textbook* site provides 25 copiable options including a Venn diagram, a fishbone planner, a semantic map, and a vocabulary concept map, among others. Links are available on this chapter's companion page: <https://jeremyhyler40.com/science-and-literacy/chapter-2/>.

No matter what the usage of the science notebooks is in your classroom, the same uses should be repeated often. In my experience, notebooks only work if they are used regularly. Students need to get used to the feel and the freedom of their notebooks, and it can take some time to establish a routine that involves notebooking. It often requires five or six entries for students to relax and start to see the value of their science notebooks. Students need to be able to see their personal growth, which cannot happen unless there are multiple entries. As a teacher, I often emphasize the value of a whole product over a specific entry — it's the whole experience that counts, and it's perfectly fine if some entries are "not the best."

To help get started, the instructions I provide my students about notebooking at the beginning of a field biology class on Beaver Island is shown in the Sidebar 2.6 below.

Sidebar 2.6 Science Notebooking Instructions for a Field Biology Class

Science Notebooking

Welcome to Field Ecology, Bio100Z!

For this course, you'll be keeping a notebook (which we provide) that documents what is happening in the classroom (lecture notes), in the field (notes and data collection), and during your off-hours (reflections about the day and observations). Some of the notebook entries will be answers to specific questions, while others will be in a more free-format. Some entries can be written, but you are encouraged to draw, sketch, press — all is welcomed!

Your notebook will be graded twice: during the midterm exam, and during the final exam. It is worth a total of 200 points (about 1/3 of the course's total points), with a 100 points dedicated to the field entries, and 100 points for other entries. It replaces lab reports or other types of worksheets. You'll get to take it home. While I'll appreciate your artistic abilities, this is not what a science notebook is about: you will not receive fewer points if you cannot draw well, nor will you receive a higher score if you are a proficient artist. Instead, I will look for accuracy, completeness, effort, connections made and growth when grading your notebooks.

Reflections are highly encouraged for your notebook, and your journal is, as such, a fairly personal space. That being said, keep in mind that it is in no way a diary, that I will read all entries, and that you'll be asked to share some entries of your choice with others, once in a while.

And to get in the spirits of things, we're going to start right now! While you're on the ferry to the island, take some time to reflect and answer the following 5 questions. Your answers do not need to be particularly long, but can be!

1. Why are you taking this course?
2. What do you expect this course will be about?
3. What are you most excited about for this course?
4. What are you most worried about for this course?
5. Isolate yourself, and take a moment to look at Lake Michigan for at least five minutes uninterrupted — describe what you see (try to be specific) and any thoughts you had about Lake Michigan.

Make sure to put your name in the front of your notebook, and to date your entries, so that I can find them later!

I look forward to seeing you all tonight!

Dr. Pangle

How to Assess Science Notebooks

As it has probably become clear in this chapter, the work that students do in their notebooks is, by its very nature, formative. And, returning to Wendy Ward Hoffer's work, we are reminded that "[a]ssessment *for* learning ensures that all of a teacher's good effort is not for naught" (p. 60, emphasis in original) and that "[a]ssessment, when done well, is part of learning" (p. 78).

For Jeremy, when it comes to science, notebooking is very much about the process, both for writing and drawing. He looks at entries to assess whether students are making an honest, intellectual effort at explaining scientific ideas both through words and images, and to see what misconceptions they might still hold. Table 2.2 is the rubric Jeremy uses with his students.

Students are assessed twice during a nine-week marking period. Whereas a focus on scientific vocabulary, the use of claims, evidence, and reasoning, and some complexity in sentence structure are all part of what he looks for in a students' writing, he does not withhold points for small spelling or grammar errors. Similarly, with the drawings, being an incredible artist is not required, just a good faith effort with some labels to help articulate what is being represented on the page.

For Wiline, she works to assess science notebooks as a whole product rather than as specific entries. This then lends itself to a pass/fail type of assessment, where effort is rewarded, and she pays attention to whether students were accurate in their representation, clearly labeling diagrams and making connections to concepts that were previously studied. For her field

Table 2.2 Jeremy's Science Notebook Rubric

Components	Exceeds Expectations (4pts)	Meets Expectations (3pts)	Needs More Work Please (2pts)
Knowledge of Concepts	Notebook is quite accurate and demonstrates concept understanding.	Scientific principles are not well understood. May contain errors in thinking.	Notebook content lacks detail and shows little understanding.
Neatness	Majority of notebook entries are neat and easy to read.	Most notebook entries are easy to read.	Notebook entries are messy and difficult to read.
Scientific Vocabulary	Student uses specific vocabulary and data to accurately communicate experiment or activity.	Student uses vocabulary and data to accurately communicate experiment or activity.	Student uses vocabulary or data inaccurately to communicate experiment or activity.
Record Keeping	Majority of drawings and written explanations describe observations clearly. Complete sentences are almost always used. Drawings are labeled, colored and include details.	Most drawings and written explanations describe observations clearly. Most observations use complete sentences. It is easy to know which experiment is being observed. Drawings are colored and labeled.	Few drawings and written explanations describe observations clearly. Incomplete sentences are used. It is hard to know which experiment the observations are about. Drawings are not clear or labeled.
Organization	Almost all of the notebook entries are easy to follow. Numbers, bullet points and spaces are almost always used to separate different observations and answers.	Most notebook entries are organized and easy to follow. Few are not. Numbers, bullet points and spaces are usually used to separate different observations and answers.	Notebook entries are not organized and very difficult to follow. Numbers, bullet points and spaces are rarely used to separate different observations and answers.
			Total:_____

Table 2.3 Wiline's Notebook Entry Checklist for Final Assessment

<STUDENT NAME HERE>	Pass/Fail
Day 1: Friday	
Initial entries on ferry ride	
Day 2: Saturday	
Lecture what is Ecology	
Walk on trail at station (look at levels)	
Lecture how we study ecology	
Lecture Behavioral ecology	
Chipmunk lab data and notes	
Free entry for Saturday	
Day 3: Sunday	
Free entry for Sunday	
Lecture notes on hyena case study	
Behavior lab with answers to 4 questions	
Day 4: Monday	
Wake up review	
Lecture on life history	
Life history monarch/milkweed lab	
Lecture on population ecology	
Snake lab — measuring dispersion	
Free entry for Monday	
Day 5: Tuesday	
Wake up review	
Lecture on community species interaction	
Field trip — notes on 2 posters, 2 people and invasive species	
Lecture on community ecology food web	
Lab notes on pollination ecology	
Free entry for Tuesday	
Day 6: Wednesday	
Mid-class reflection	
Lecture invasive species	
Pulling spotted knapweed lab	
Garbage pit evening outing and trail cams	

biology class, mentioned earlier, this spreadsheet/checklist simply lists all the entries that she has asked students to do and — as long as they have made a legitimate effort at creating an entry for each one — they earn a passing grade. An entry that is insufficient or missing earns a failing grade. See Table 2.3, a checklist from her 10-day intensive biology course, offered on Beaver Island, as one example.

<STUDENT NAME HERE>	Pass/Fail
Day 7: Thursday	
Lecture community dynamics	
Invertebrate diversity lab with answers to 2 questions	
Lecture primary production	
Free entry for Thursday	
Day 7: Friday	
Wake-up review notes	
Lecture secondary production	
Dune succession lab	
Productivity lab with answers to 2 questions	
Lecture nutrient cycling	
Bog outing entry	
Day 8: Saturday	
Lecture climate change	
Lecture conservation	
Boat outing entry	
Ecological footprint and reflections	
Day 9: Sunday	
Free entry for Sunday — reflection on the class	
Milkweed wrap up lab — look for patterns	
Notes on all presentations (9)	
Review session	
Additional Criteria	
On-going notes and data for final project	
Notes on interviews on career path of 5 people	
Clarity and ease of reading	
Connections made to concepts from class	
Use of multiple media (incl sketches, drawings, etc.)	
Personalized, unique approach to notebooking	

On occasion, she will collect a specific notebook entry that will receive a more formal assessment. With these entries, she provides more detailed feedback, annotating segments that could use elaboration or where a potential connection could be made. Similarly, if there is a mistake or misunderstanding, she corrects that, too. In this case, a rubric for the assignment would include elements such as:

- employing precise vocabulary and using it accurately;
- connecting to course content, accurately, and in detail;

- drawing conclusions based on evidence; and
- including items like drawings and diagrams that are labelled and logical.

Given these ideas, the power of notebooks is that students already have the materials needed; there's nothing fancy that they demand, and students begin to take ownership of their notebooks — and the ideas contained within them — as the semester or school year progresses. As a space to begin their inquiry and document the process of learning, their notebooks prepare them well for the kinds of work we ask them to do when representing their thinking visually through infographics (Chapter 3) and in the process of modelling instruction (Chapter 4). With the groundwork laid in their notebooks, we move students to these additional forms of writing-to-learn.

Next Steps with Science Notebooks

Even though both Wiline and Jeremy prefer using actual, paper notebooks with their students, digital options are emerging. As the ISTE Standard for Students suggest, they can be "knowledge constructors" who "curate information from digital resources using a variety of tools and methods to create collections of artifacts that demonstrate meaningful connections or conclusions" (3c) and "build knowledge by actively exploring real-world issues and problems, developing ideas and theories and pursuing answers and solutions" (3d). In this sense, we can invite them to take their notebook entries, whether analog or digital to begin with, and extend their ideas with digital tools.

In addition to the ideas mentioned above, Common Sense Education has created a list of 13 tools for engaging in citizen science, including ones that could be integrated into the process of notebooking (n.d.). First, the Science Journal by Google, available for free, is described by Common Sense Education as a way to "[t]urn your phone into a lab sensor to collect and analyze data." Similarly, The PocketLab, created by Myriad Sensors, can be installed on a mobile device, but does require at least one of any number of additional sensors, and kits ranging from less than $10 to over $150. Finally, Zooniverse, "the world's largest and most popular platform for people-powered research" (n.d.), offers projects across a variety of fields including biology, nature, physics, space, and medicine, among other topics in the humanities and social sciences. As a reminder, for direct links to all these resources, please visit this chapter's companion page on the website <https://jeremyhyler40.com/science-and-literacy/chapter-2/>.

When we invite students to begin their inquiry in notebooks, there really is no telling where their ideas might go. And, this is the nature of curiosity; we want them to use writing as a way to explore and discover. With this in mind, we now turn our attention to a combination of literacy and numeracy, where more insights can be drawn from examining data and representing it through graphs and infographics.

3

Visual Explanations with Infographics

For many students, looking at a textbook and seeing the countless numbers of visuals can be overwhelming. Yet, we know that visuals are essential and, we hope, were intentionally designed by the textbook authors. Combine those visuals from more formal academic sources with the kinds of charts that students see in the news and through social media outlets, and we must ensure that they are capable of interpreting — and creating — their own visuals in a clear, compelling manner. In a broader sense, we also know that students are living in a visual culture. As Martix and Hodson note, "the creation of pictorial representations of written arguments requires that students engage in important critical analysis of the material that they are learning" (2014, p. 19).

One of the greatest challenges facing science teachers is getting students to understand data that are placed in front of them. Furthermore, it is rather difficult for students to make comparisons between two different sets of data. For instance, when students look at data from two different runs of the same experiment, they may not know how to take the combined data set and organize it with a table (or, they can do it, but they can't explain it). Part of the work of combining aspects of science and literacy, then, is focusing on numeracy and graphic design, too.

Thus, we argue that students can and should learn how to better interpret data through infographics. Infographics are visual representations of data or information that are meant to display information in a quick and easy way for clear interpretation. They are particularly helpful for those students that are visual learners. As Troy and his colleague Kristen Hawley Turner describe them,

Infographics have become an increasingly popular means of expressing information and making arguments. Perhaps because of their visual appeal, perhaps because they are so easy to "like," "retweet," or "pin," infographics or a study part of the content readers consume online.

2016, p. 60

Anyone who spends any time on social media is bound to have seen an infographic, and simply searching for "infographic examples" will yield millions of hits. The visual components of infographics are made to be appealing to readers while providing information about a specific topic, and for a specific purpose (see Sidebar 3.1 for a strategy to help students discern this information).

Infographics should help students, quite literally, see the information that is trying to be relayed. Too many times, teachers have students create infographics and they are nothing more than a glorified poster where there really is not any data (let alone data that have been collected by the students themselves). This doesn't mean that the teacher's intention for integrating infographics was bad. Yet, we believe that students gathering and then representing data is a crucial skill.

In other words, just because we are asking students to use Piktochart, Infogram, Easelly, Visme, Animaker (all linked from the companion page) or other similar types of tools, that alone does not make the product an

Sidebar 3.1 A Visual Thinking Strategy Useful for Analyzing Infographics

At the heart of it all, we need to remember that any kind of visual we use in science must be purposeful. And, if it isn't evident by what we have stated so far in this book, we all agree that the role of data in scientific arguments is to provide evidence.

If data are not organized and presented effectively, it can lead to an inaccurate argument and, in the end, a misunderstanding of the data being presented or, worse yet, a scientific misconception. In this sense, teaching students to be visually literate and numerate are both tasks that fall to science teachers, and both fit in with a larger goal of teaching them to be scientific writers.

Recently, Wiline started to use Visual Teaching Strategies (link available on companion page) to introduce graphs, especially in non-major courses. They offer great resources, although a subscription is required to get to most of them. One simple method (Housen & Yenawine, n.d.), available for free and developed originally to talk about art, invites students to discuss a visual with three simple, yet essential questions:

1. What's going on in this image?
2. What do you see that makes you say that?
3. What more can you find?

infographic. Each of these tools allows users to either input or import data, sometimes directly from a Google Spreadsheet. With this capability, we can have students review major data sets and sort out useful subsets for their own infographics, and we can have them take the results of their own surveys and transform those as well.

Having said all this, for this particular chapter, we want to help educators realize infographics are about the data, especially the data that students are collecting themselves. Throughout the chapter, we will explore how Jeremy and Wiline each use strategies for teaching visual literacy with specific attention to infographics.

What Infographics Are

For a concise definition of infographics, we turn again to Troy with Kristen Turner in their book *Argument in the Real World*. Here, in a brief summary, they point out that "infographics are a combination of words, numbers, and visual elements." This is true, "[y]et," they continue, "this description oversimplifies the way that writers think about composing this type of digital text" (62). They argue that infographics must have a compelling story, draw data from a reliable source, need to be aesthetically pleasing, and, in keeping with the spirit of sharing on social media, be easy to share. To draw from a leading scholar on visual media and the representation of data, Edward Tufte makes the case in his classic work *The Visual Display of Quantitative Information* (2001) that

> [W]hat is to be sought in designs for the display of information is the clear portrayal of complexity. Not the complication of the simple; rather the task of the designer is to give visual access to the subtle and the difficult—that is, revelation of the complex. (191)

When it comes to infographics, in other words, the goal is to keep it simple.

So, we see the design of graphs, as well as more elaborate infographics, as a form of writing, as shown in our MAPS for designing graphics and infographics (Table 3.1).

A few additional teaching resources that we find helpful include math teacher Kelly Turner's "Graph of the Week," an assignment in which Turner believes "[s]tudents are given the opportunity to communicate their critical thinking and analysis through writing and classroom discussion" (K. Turner, n.d.). She uses a consistent format week-to-week, asking her students to first make some observations and then ask some questions. To observe, she suggests that they identify the topic of the graph, describe what the x- and y-axes

54 ◆ Visual Explanations with Infographics

Table 3.1 The MAPS for Writing with Graphs and Infographics

Mode	Simple hand-drawn graphs or more formalized graphs and infographics made with an app, program, or website with spreadsheet and design functions.
Media	Building on both literacy and numeracy skills, graphs include numerical data represented in a visual form, with appropriate descriptors, colors, and fonts.
	As a more stylized version of graphs, infographics may also include other design elements such as banners, icons, images, and elements to capture a viewer's attention.
Audience	For hand-drawn graphs, students are generally creating materials for an audience of themselves, their peers, and their instructor.
	For more formal graphs and infographics, these materials can be shared with wider audiences through blogs, social media, presentations, or other means.
Purpose	Building on both literacy and numeracy skills, asking students to create hand-drawn graphs as a formative assessment — as well as more formal graphs and infographics for summative assessments — helps them evaluate and synthesize data, representing key elements in a significant manner.
Situation	Graphs can be included in nearly any lesson that is connected with data.
	As a short-term, informal opportunity, students can be asked to sketch a graph during or immediately after an experiment, with an existing set of data they are examining, or as part of a modeling/whiteboarding exercise (see Chapter 4).
	As part of a long-term, more formal opportunity, students can be asked to collect data over time, adding to a graph over subsequent iterations of an experiment or observation. With an infographic, they can represent their data to an outside audience through a visual argument, employing effective design techniques to accentuate their data and drive home their message.
	Students can also then write descriptive analyses of the graphs that they have created, articulating key design decisions and reflecting on what they have learned in the process.

Sidebar 3.2 Journalist Mona Chalabi's Work

Mona Chalabi, a journalist for *The Guardian*, creates hand drawn graphs (and short videos) that represent data about contemporary social issues, targeted directly at viewers of Instagram and Twitter. They are both witty and biting, making commentary on many social justice issues such as racial discrimination in wages, incarceration, and housing, vaccination rates, immigration, climate change, and wealth inequality. Definitely preview her work before using it with your students, but do take the time to appreciate how she designs her images. Follow links to her work on the book's companion page, <jeremyhyler40.com/science-and-literacy/>.

For instance, as we were putting the finishing touches on this manuscript, in September 2019, a number of reports were being released about the effects of e-cigarettes on health, tying the act of vaping to recent deaths (Richtel & Kaplan, 2019). Chalabi created a timely graph, released on September 11, 2019, titled "High School Students Who Smoke Cigarettes or E-Cigarettes" (Chalabi, 2019). On the x-axis, the horizontal images of both a cigarette and e-cigarette are on the left, and the dates along the bottom of the graph then start at the end of each image, stretching from 2011 to 2018, with one hash mark for each year. Then, on the y-axis, the percentages rise vertically from 0% to 25%. In 2011, the image of the cigarette is leveled out at about 16% and the "smoke" coming from the end of it continues to drop as the data moves toward 2018, with a slight uptick at the end. In 2011, the e-cigarette is at about 1% and the line of "vapor" rises quickly, with one drop in 2016 to a high of about 22% in 2018. Through visual metaphor, she shows how the decline of smoking and the rise of vaping are intertwined.

represent, and to make some general observations and predictions about the data. She then asks them to consider upward/downward trends, sudden spikes or dips, what items are specifically being compared, and what inferences can be drawn. Turner has an archive of weekly graphs going back to 2013, all available on her website, and linked to this chapter's companion page. Another interesting, though slightly edgy (and not always appropriate for school), data visualization specialist is Mona Chalabi, a journalist for *The Guardian* (see Sidebar 3.2).

And, as one last, yet certainly not least strategy, a simple web image search for "infographics ___" where you add the topic to the search query will yield hundreds, if not thousands of hits. Finding the infographics, to be honest, is not the difficult part. Having a clear rationale for students when interpreting and then creating them, however, is.

Why Use Infographics

This will come as no surprise, but it is worth noting that many students will fall back on what they know, especially when it comes to making their visuals. That is, many of them may be comfortable with making a general bar chart, but they may not be able to make a stacked bar chart. Moreover, they may not fully understand why a certain type of graphic is a strong representation of the data. Throughout this chapter, we provide some detailed descriptions about when, how, and why to use particular kind of graphics in particular ways because students are often missing the fact that certain types of visuals display information in ways that are better than others.

Also, before going too much further, we need to clarify a few terms. Often, we hear "charts" and "graphs" being used interchangeably. To borrow from a brief article in *Sciencing* on "The Difference Between Charts & Graphs," that difference is subtle, but useful. Author Karen Blaettler suggests that "[c]harts present information in graphs, diagrams and/or tables. Graphs comprise a specific type of chart, showing the relationships between mathematical data" (2018). In some ways, we wonder if this broad genre of visuals should be called "infocharts," then, instead of infographics. But, alas, they are not.

So, for now, we need to remind students that graphs — which could be thought of more thoroughly as *graphical representations of data* — must be something more than just a table of numbers; bar graphs, line graphs, pie graphs, and scatter plots all represent items in the broad category of "graphs."

In fact, Wiline notes that she does not often use the term "chart" when talking with her undergraduate biology students, in an effort to keep the ideas separate. Charts, again, are *all* the visual elements. Graphs are about relationships between the numbers.

With that, the main idea is that graphics — as well as infographics — are everywhere. As an exercise, it is always important that our students learn how to view and interpret these visuals. At a deeper level, we also need to encourage them to do more as they become data analysts and visual designers themselves, creating infographics that represent their inquiry in science.

How to Teach with Infographics

Deciding on the specific times and places in our crowded curriculum to take extra minutes, even entire class sessions, to devote to the process of building data into tables and graphs (and, subsequently, moving those initial representations of the data into an aesthetically-pleasing, shareable infographic) is, well, a challenge. Like all choices that we make in teaching, the amount of time that we devote to allowing students to create an initial graph and, in turn, the full infographic, is one that depends on the time of the school year, the unit in which students are learning, and their level of interest. Wiline uses graphs and pictures from day one, and asks her students to begin creating them immediately.

In whatever time we are able to allow for moving from initial data collection all the way through to a final infographic, one question that Wiline uses with her students helps guide their work is this: *what is the story that your data is telling you?* By returning to this question, she reminds her students that understanding the patterns in the data and making them clear for other scientists (or students) is critical. In other words, she sees incorporating infographics as an essential part of teaching science. Put another way, she can talk about a scientific concept for 10 minutes, trying to get students to visualize the concept, or she can show infographics about the concept that conveys the message in a more aesthetically-pleasing and accurate way, all in just a few moments. In this case, the cliché that "a picture (or infographic) is worth a thousand words" does ring true.

As they begin looking at infographics, essentially consuming the work that has been shared from someone else's inquiry, Wiline invites them to think rhetorically and strategically about how to represent their own data. She poses questions like this to get them moving along (Sidebar 3.3). These questions provide the perfect segue into a discussion on data representation to help students think through the decisions that the graph's designer made, thus informing their own thinking as they prepare to make their own.

Sidebar 3.3 Questions to Prompt Student Thinking Around Infographics

- What trends do you notice?
- Why did you set up the graph in the way that you did?
- How does the graph help you organize your data so you can build evidence into an argument?
- Why is my graph one that best represents what is going on?
- What would happen if …
    - … we change the grouping of the data?
    - … the data were continuous? Or categorical?
    - … we couldn't use color in our graph?

Jeremy's Perspective: Introducing Infographics to Students

For me, the introduction of infographics takes place in both the English and science classroom. I see infographics as a type of writing because there is a story to be told as well as an argument to be made. The story/argument can vary, showing what a student has learned across an entire unit or simply something that is created at the end of a single lab. Both aspects of the infographic design process are important, and I invite students to focus their attention on each, but to varying degrees.

In terms of story, I want the infographic, like different genres of writing, to have its own style. That style is illustrated through various visual elements such as pictures, graphs, and diagrams, as well as design elements such as colors, fonts, and backgrounds. And, while style is important, I usually only weigh this aspect of the infographic at about 25% of the overall grade. In other words, while creating a visually pleasing infographic is important, the main focus needs to be on the argument being presented from evidence.

Thus, in terms of argument, I point out to my students that it is important that we are providing readers accurate information that was gathered through a scientific process. Similar to a traditional expository essay, any information provided should be factual and not based on student opinion. Depending on the claim they may be making, the visual elements can provide evidence for the claim given by the student. In other words, representing their data in an accurate and complete manner is more important than style, usually weighing about 75% of a student's grade on an infographic.

By the time I walk through all the ways infographics are related to writing with my students, and emphasize the value of both story and argument, it is time to shift to our first practice assignment that focuses on reading and interpreting infographics so they have an example to draw from. I use a lesson, "Using Graphs for Writing-to-Learn," that was originally designed by Kathy Kurtze, a Chippewa River Writing Project colleague, and has since been adapted by Troy. Then, for my class, I have adapted it further. You can

Sidebar 3.4 Using Graphs for Writing-to-Learn (Originally created by Kathy Kurtze and adapted by Troy Hicks)

> Graphs found in textbooks, newspapers, and online can be used to create meaningful opportunities for students to write. By answering a standard set of questions in complete sentences, students are able to show their understanding of a graphic in a paragraph format. Once students become accustomed to the set of questions, they become more adept at formulating their answers, and are more likely to be able to transfer the format to other situations. Also, they can take this initial set of sentence templates and then mix, match, and otherwise revise to make more complex, nuanced summaries.
>
> **Directions**
>
> Study the graph. Analyze what is being represented. Determine what is being studied, or what question is being asked and answered. You can do all of that by addressing the following prompts in complete sentences. Feel free to sprinkle in more details from any other research you've done or background info you may have.
>
> - State the title of the graph.
> - Describe the picture/shape of graph.
> - Summarize the information given in the graph.
> - Cite specific statistics and results included in the graph.
> - Name the source of the information.
> - Offer your opinion of the results.
> - Explain how the information relates to you or the world around you.

find a template for this lesson on the book's companion page for this chapter <https://jeremyhyler40.com/science-and-literacy/chapter-3/>. The lesson asks students to write a paragraph-long summary, as shown in Sidebar 3.4.

As an example of this lesson, I have my students refer to an infographic (Image 3.1) from the United States Department of Agriculture (2013). For further reference, the USDA has an entire gallery of infographics on Flickr (linked on our companion page), and there are many other places to find useful infographics through a general web search. Figure 3.1, "Bringing the Farm to School," is one example that I have used with my students, as it is highly relevant to our rural community and the environmental aspects of science that we study.

To begin, I pass out a copy of the infographic to each student and display it on the projector. I ask the students to record in their science notebooks what they notice. I simply have them make a list with bullet points. Then, I move into more specifics with my students. After they have made their list, I ask them to describe the images that they see and why are they relevant to the infographic. From there, the students are directed to summarize in 2–3 sentences what information is in the graph. Below is a typed summary from a 7th grade student this past school year.

Figure 3.1 USDA Infographic: "Bringing the Farm to School"

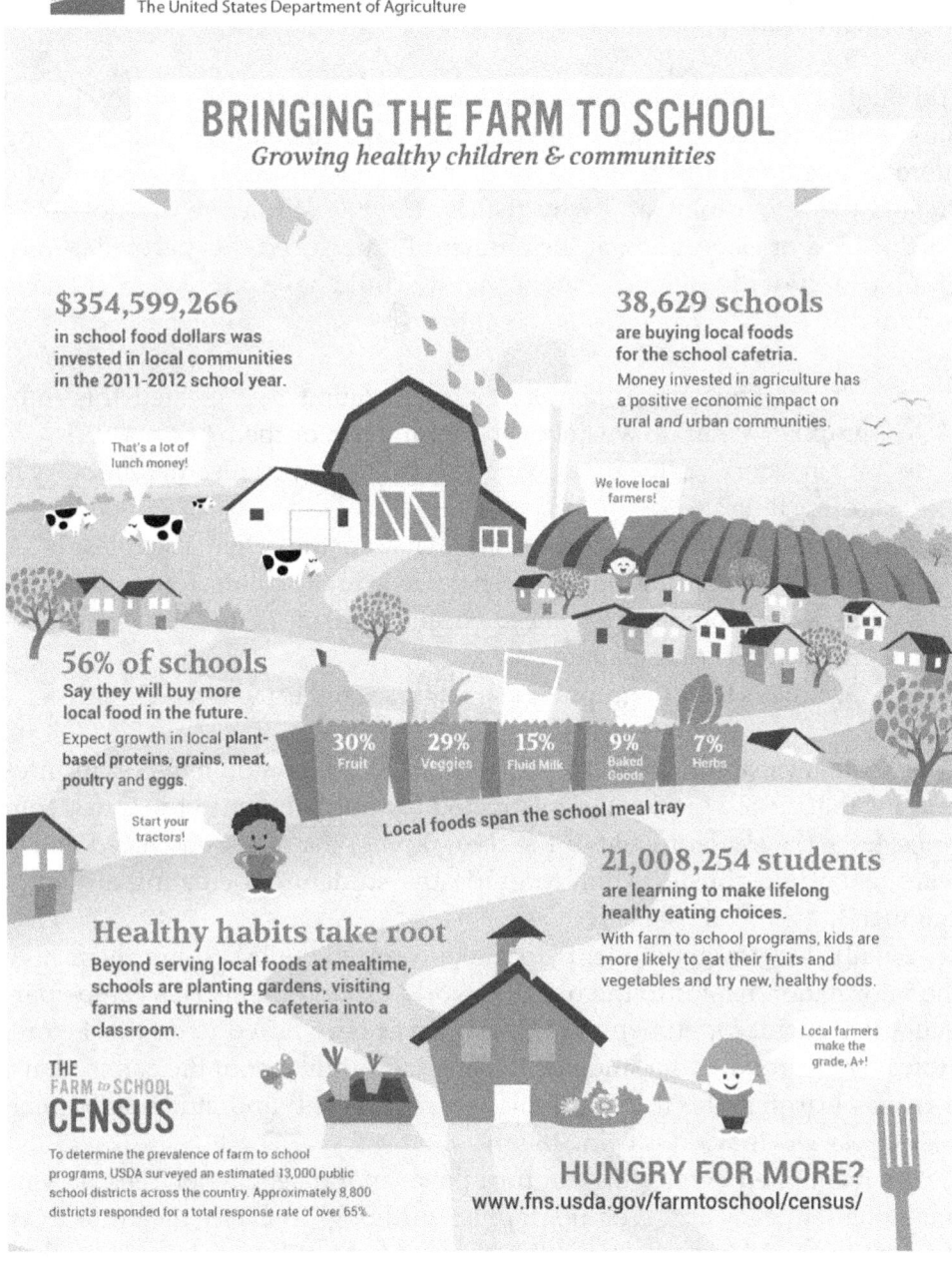

> *The infographic is trying to get people to understand what it takes to get children to eat healthy. It breaks down the number of schools, students, and communities that are choosing to change their habits for eating healthier.*

Although this student does a decent job of summarizing the poster, I want all students to dive deeper into what this particular infographic is, being more detailed about it. In addition, I may even have students discuss who created the infographic and why that is relevant. While we did not strive for this kind of elaboration at the moment I introduced this particular infographic to students, other questions that we could have pursued might have included:

- Who is the person, people, or organization that created this infographic? What do we know about him, her, or them?
- What "story" is the data trying to tell us? Do you feel that the story is compelling?
- Where does the data come from? What are the original sources?
- From a visual standpoint, in what ways do the colors, fonts, and overall design combine to be aesthetically pleasing? Are these visual elements literal or metaphorical?
- Overall, is the infographic clear, concise, and easy to read?

For a student example, Figure 3.2.1 and 3.2.2 are components of an infographic that an 8th grade student created for our Salmon in the Classroom project. As the student created the infographic, the questions listed above were taken into consideration to guide the students in creating a quality product.

Finally, I try to get students to make connections by asking them how the information relates to them or the world around them. This can be particularly difficult for students who have never been asked to do this before. However, by guiding the students to pause and think about the connections, it makes learning about infographics more relevant, and students become vested learners in what is being taught.

As mentioned before, we can turn pages in a magazine or click on various internet pages and see infographics. Although they are meant to relay information quickly and clearly, the question may still remain on how they can help students. For my students, we find it beneficial that they use infographics for representing data that they gather in labs, especially data for comparative analysis.

Visual Explanations with Infographics ◆ 61

Figures 3.2.1 Student-Created Infographic from the Salmon in the Classroom Project

Fish Population in Fish Creek.

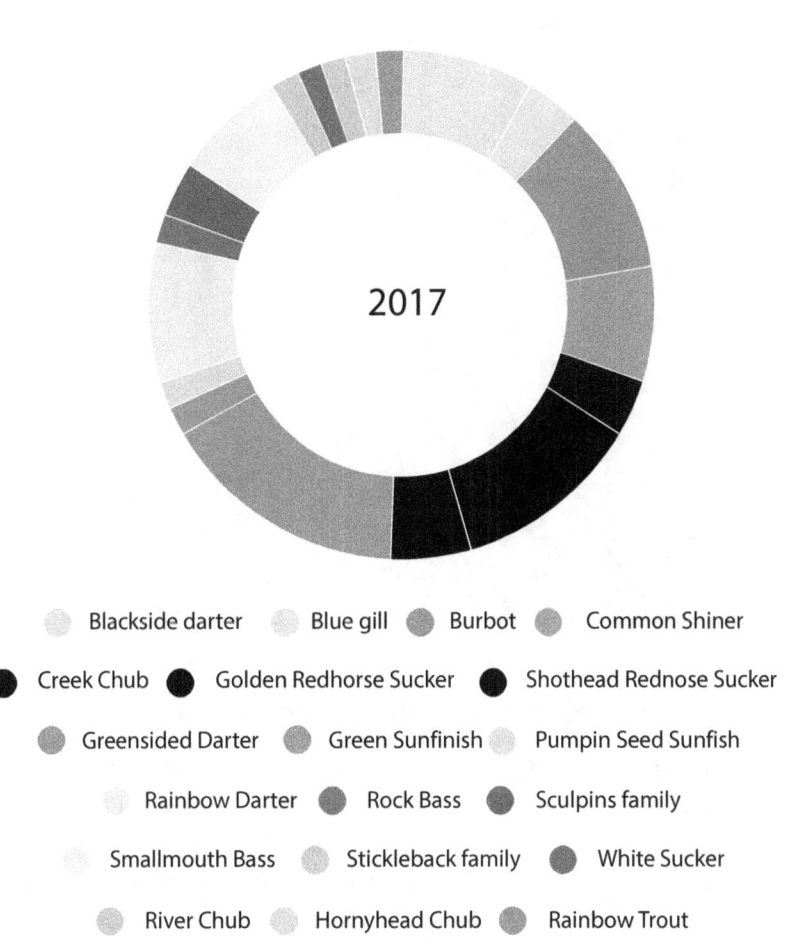

Salmon in the Classroom - 8th Grade - Fulton Middle School

2017

- Blackside darter
- Blue gill
- Burbot
- Common Shiner
- Creek Chub
- Golden Redhorse Sucker
- Shothead Rednose Sucker
- Greensided Darter
- Green Sunfinish
- Pumpin Seed Sunfish
- Rainbow Darter
- Rock Bass
- Sculpins family
- Smallmouth Bass
- Stickleback family
- White Sucker
- River Chub
- Hornyhead Chub
- Rainbow Trout

Over 61

Fish that were documented

Figures 3.2.2 Student-Created Infographic from the Salmon in the Classroom Project

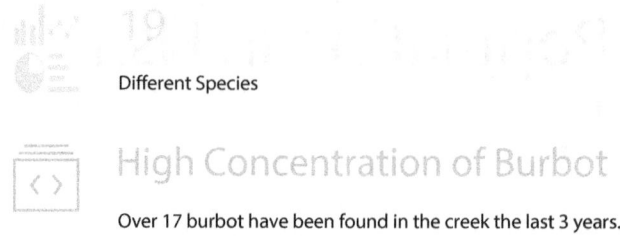

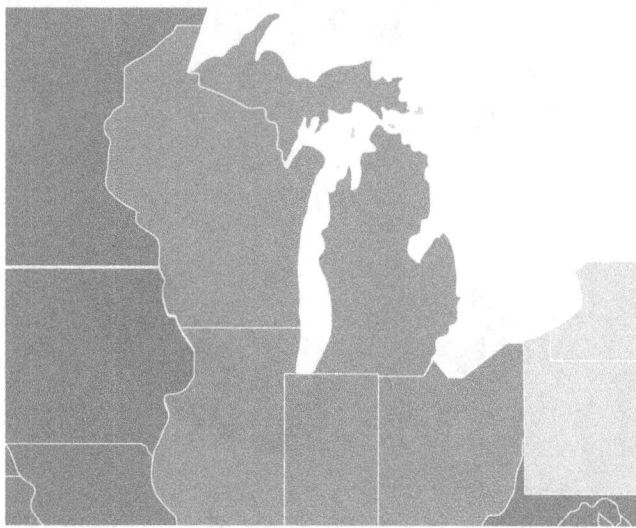

Wiline's Perspective: A Deeper Dive into Creating Tables and Other Graphs

How we represent data matters. Yet, we often skip the discussion about what type of representation we should use for specific dataset. This segment of the chapter describes an activity that I practice to target this specific question: how do we represent data? In fact, data can be represented visually in a number of ways. Each method has its advantages and emphasizes a different aspect of data, and I try to explore that big idea in this lesson.

For this example, we are going to examine the fat contents of cookie dough and how cookie weight, diameter, height, and moisture are each affected in the process of baking, based specifically on fat levels. Or, in plain English, we are looking at whether higher levels of fat might make cookies bigger, more moist, and ultimately more tasty. These data are actually drawn from an article in The *Journal of Food Engineering*, and help provide some straightforward numbers with a real-life example, thus making it a good example for my students (Pareyt et al., 2009).

By making multiple versions of a graph with the same data, beginning with a table, I believe that a lesson like this is worth consideration for students of all levels — and sometimes for ourselves! The approach to looking at a single set of data — and leading students through an iterative process of analysis and graph design — allows them to understand the affordances and constraints of their design choices. For instance, the Visme blog describes 44 different types of graphs ranging from the typical bar, line, and pie graphs to mosaic charts and population pyramids (Lile, n.d.). Let's examine how I introduce this lesson to my undergraduates.

Step 1: Organizing Data into Tables

Tables are a great first step to organize data systematically before displaying the data in any kind of graph. A table — quite simply defined by Merriam Webster as "a systematic arrangement of data usually in rows and columns for ready reference" — is a great way to organize a large amount of data quickly (Merriam-Webster Online Dictionary, n.d.-b). Though there are some very nice templates in programs like Excel, Numbers, and Google Sheets, the basic design of a table is not visually appealing. It can be hard to stare at a table and see patterns emerge.

Table 3.2 is an example of a simple table drawn from the article mentioned above, Pareyt et al.'s "The Role of Sugar and Fat in Sugar-Snap Cookies: Structural and Textural Properties." Before we dig into the actual table, it's important to remember that any dataset can be used in such an activity, provided it remains simple (no more than four to six variables) and really small; there's no need for an extensive dataset, which would take students time to process and might be discouraging as they make mistakes.

Table 3.2 contains only a few rows and columns. Notice that the table is organized based on the fat content of cookies (the first column), and its effects on four different aspects of cookies (the next four columns). As the data are presented to students, we can encourage students to organize the

Table 3.2 Table Summarizing Data from "The Role of Sugar and Fat in Sugar-Snap Cookies: Structural and Textural Properties"

Dough fat level (%)	Cookie weight (g)	Cookie diameter (mm)	Cookie height (mm)	Cookie moisture (%)
15.8	23.5	90.1	7.1	2.7
13.1	22.8	86.8	8.2	2.1
10.2	24.6	82.8	8.9	3.5
8.7	25.2	79.6	9.9	3.9

rows based on the data points in the first column: by listing the items in order of decreasing fat levels in the dough (the first column), this brings some focus to the data. Yet, despite this organization, it would be hard to figure out in just a few seconds of examining the table how fat is affecting the cookies' weights, diameters, heights and moisture. Moreover, it is very difficult to discern how some of these characteristics interact with each other. In other words, there is no way to know what is tasty and appealing in these cookies!

At this point, I ask students to consider the following questions:

- What are some advantages of organizing data into tables? Disadvantages?
- Are some types of data better suited for presentation in a table? Which kinds? Why?
- What are some of the limitations of tables in terms of identifying patterns within a variable, and between variables (interactions)?

In general, students begin to note that tables are incredibly difficult to interpret, even with only a few data points. Students often say that tables remain dull and unappealing to readers, even those that use some colors and design elements.

Step 2: Organizing Data into Graphs

Since the table format has some limitations, I then ask my students to consider how they might visually display the data using graphs. Returning again to Merriam-Webster, we discover that graphs are defined as "a diagram (such as a series of one or more points, lines, line segments, curves, or areas) that represents the variation of a variable in comparison with that of one or more other variables" (Merriam-Webster Online Dictionary, n.d.a).

Remember that I use the term "graph," almost exclusively, and not the term "chart." As a quick point of clarification, returning to the difference described by *Sciencing*, "charts" refers to a broad, umbrella term for all kinds of visuals whereas "graphs" are a "specific type of chart, showing the relationships between mathematical data." For middle school students, perhaps this distinction is too specific and technical. However, as they progress through high school and college, it is important to note that graphs, common in scientific writing, are different from charts, which can also include tables, diagrams, and other visuals. The point to consider here is that there are three primary ways to represent the data in graphs: line, pie, and bar. First, we start by creating and examining line graphs.

Sidebar 3.5 Dependent vs. Independent Variables

> Identifying the two types of variables presented in any infographic is critical to both interpretation and data representation. Students need to determine: which variable(s) is(are) the dependent variable(s)? The independent variable(s)? Even as college freshmen who should have some exposure to algebra, students struggle identifying these kinds of variables, and feeling confident about their answers.
>
> By definition, a dependent variable is a parameter of the experiment that is manipulated by the researcher, thus "dependent" on other variables. It can be referred to as "the response variable." In our example, the only data point that is actually manipulated by the researchers is the amount of fat content in the cookie dough.
>
> The other four cookie parameters (diameter, moisture, height, and weight) can be potentially affected because of changes in fat levels controlled by researchers, and as such are "independent."
>
> Wiline often reminds her students that if a variable is measured to see what happens to it in the experiment, then this variable is a dependent variable.

Step 2.1: Organizing Data into Line Graphs Considering the main data that we have for the cookies—fat content in relation to other variables—students need to determine where the variables will go in a line graph (see Sidebar 3.5 for a quick description of variables).

By identifying the dependent variable (here, the fat content of the cookie dough) and the independent variables (here, the weight, diameter, height and moisture of cookies), students can realize that they need to graph how the dependent variable (fat content) affects the independent variables (what the cookies look like). Really, by identifying the dependent and independent variables, students are mapping out the rationale of the researchers' experimental design: what was the researchers' hypothesis? What predictions did they make? Are the predictions supported by the data of the experiment? Each graph can then become "an answer" to each specific prediction, which I ask students to articulate before graphing any data:

- Prediction 1: Cookies with lots of fat will be moister than cookies with less fat.
- Prediction 2: Cookies with lots of fat will be flatter than cookies with less fat.
- Prediction 3: Cookies with lots of fat will be larger than cookies with less fat.
- Prediction 4: Cookies with lots of fat will be heavier than cookies with less fat.

Once students have laid out the rationale, graphs can be created by relating the dependent variable (the fat level of the dough) to any of the independent

Figure 3.3 The Effect of Fat on Cookie Weight

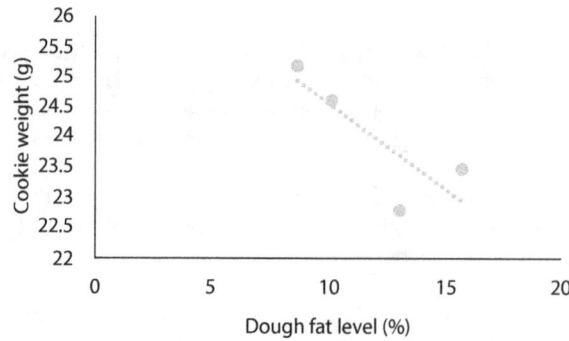

variables (here, in Figure 3.3, the cookie weight as an example), to "answer" Prediction 4. In other words, making this graph allows the students to confirm quickly whether the prediction made was, indeed, supported by the data.

In Figure 3.3, "The Effect of Fat on Cookie Weight," only two variables are represented: dough fat level (column 1 in the table) and cookie weight (column 2 in the table). Notice that in this first graph, there seems to be a strong relationship between the amount of fat in the dough and the cookie weight, with fattier cookies weighing less. This pattern is further accentuated by adding what a scientist or mathematician would call a "best fitted line," or in more common terms, a "trendline."

This graph is easy enough to create, yet (even with this limited amount of data) it might not be telling the whole story. The trendline suggests that there is a fairly strong pattern between fat level and cookie weight. This is where the first set of choices comes in, one that a software program might make for the user: if one looks carefully, while the X-axis starts at 0 and increases, the Y-axis starts at 22, and only covers a small range of values, up to 25. It results in a sort of "blowing up" of the data at this very small range by manipulating the scale of the Y-axis. This is a common misrepresentation of data, one that the National Forum on Education Statistics in their *Forum Guide to Data Ethics* warns against (National Forum on Education Statistics, 2010).

With this warning, students can see that manipulating the Y-axis is quite deceiving. Of course, this is one of the lessons we can also learn in media literacy. As Frank Baker notes, "infographics are designed by experts who employ visual literacy and art techniques," and their goal is often persuasion (2017 para. 12). In this sense, we may teach our students to look at graphs and determine how and if the scale is appropriate or, as Baker notes, if they are intentionally designed to be misleading.

To that point, Figure 3.4, "The Effect of Fat on Cookie Weight, to Scale," represents the exact same data points with axes that are scaled to go to zero.

Figure 3.4 The Effect of Fat on Cookie Weight, to Scale

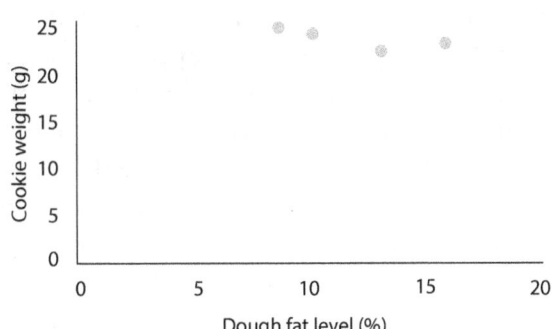

With this graph, all I have done is to scale the Y-axis (cookie weight) to go from 0g to 25g, instead of from 22g to 26g (as in the first two graphs). Now, it becomes more obvious that—while there is a relationship between the two variables—that relationship is nowhere as strong as the first graph made it seem.

But what about all the other variables? While it may be tempting to graph all parameters (weight, diameter, height and moisture) in one single graph, this is a limitation of the line graph, at least for this kind of data. In this case, all four independent variables should be graphed separately, since they have different ranges of values and, even more importantly, different units. Fundamentally, each parameter is measuring a different component about cookies. To put all of them on the same graph would be a mistake; the graph would be incoherent; sadly, this is what most spreadsheet software programs would output on an automated setting. On the other hand, if all of the data were presented in the same units, and had similar data ranges, it would be more meaningful to design a strategy for putting all of it on one graph.

It is beyond the scope of this chapter to share a full exploration of all four parameters on the effect of dough fat, though this is a task that I would ask students to do to demonstrate their understanding. Students can use such graphical exploration as actual lines of evidence to check whether their original predictions are supported by the data. In addition, having all patterns next to each other allows a qualitative comparison of the strengths of each effect.

In summary, thinking about the affordances and constraints of line graphs, we need to consider what the researchers' original experimental design was aimed at, and explore data with line graphs to follow the logic of the experiment. More importantly, we must recognize that lumping too much data into one graph without thoughtful consideration of the "story" the graph should tell is, ultimately, a bad idea. Would considering the ever-popular form of pie graphs be a better solution?

Step 2.2: Organizing Data into Pie Graphs Another popular style, the pie graph, is useful in showing how much each group contributes to making a whole. They are easy to interpret, but only apply to a very specific type of data. They can be very effective at showing the relative proportion of each species to the overall sampled community, or how many people might hold specific opinions about a topic out of all people asked.

For the cookie data, if we attempt to pie graph the same relationships we did for the line graphs above, we would generate a meaningless pie graph. Why is a pie graph a poor way to represent this particular kind of data? First, there was not a whole group sampled. There just happened to be four different types of cookies, not an entire population of cookies. Nothing adds up to 100%, so no matter how we look at the data from Table 3.1, there really is no way to graph it as a pie in a way that has any coherence.

An interesting exercise here would be to think about what kind of data students would need to have in their original data collection to generate a meaningful pie chart. For instance, if 100 people were asked to taste each of the four types of cookies (each varying in their amount of fat content), and then those people were to report their favorite type of cookie, things could get interesting; in this case, a pie graph could quickly reveal which kind of cookie that people like best, but that is not the kind of data that we have available here.

Pie graphs, like line graphs, have their advantages and disadvantages. Thinking about pie graphs, we need to consider their visual appeal and the ease with which they can be constructed, while remembering their narrow usage. As a general rule, remember that pie graphs are best for showing a distribution of sub-populations within an entire population. Thus, it is not the solution for this kind of data in our cookie study. We look next to bar graphs.

Step 2.3: Organizing Data into Bar Graphs Though we could also consider box whisker graphs, surface graphs, or other types of clustered graphs, I usually end the lesson with having students represent their data in bar graphs. Just as line and pie graphs have their distinct features and purposes, bar graphs have an advantage, too: they are a great visual tool to represent averaged data by comparing groups to each other. Similar to pie graphs, they use a specific type of data, and may not always be appropriate.

Using the data from the table about cookie fat levels, students can create a bar graph that represents the changes in cookie moisture based on the cookies' fat levels. While it would be tempting to let each dough fat level have its own column, to use bar graphs appropriately, we need to lump the data points into groups; specifically, we can create a "Low fat" and a "High fat" group, and compare the two (see Sidebar 3.6 for a discussion creating categories in bar graphs). Students can average the moisture levels of the two types of low-fat cookies (3.5 and 3.9, for an average of 3.7) and of the two high-fat cookies (2.1

Figure 3.5 The Effect of Fat on Cookie Weight, as a Bar Graph

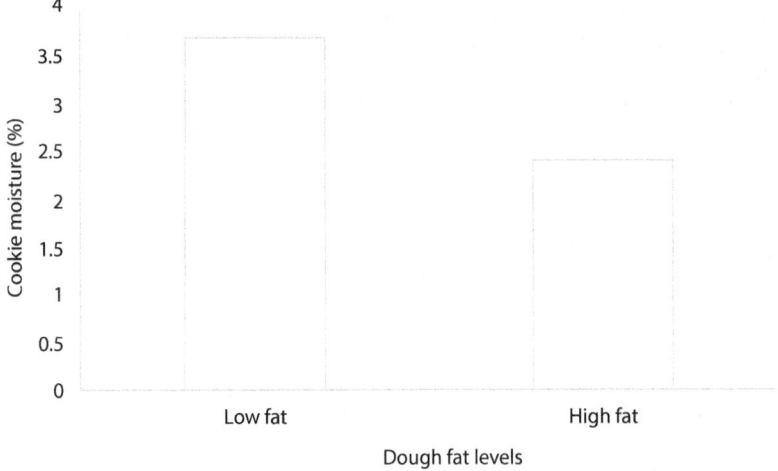

and 2.7, for an average of 2.4), and represent the two groups, one for each bar. In this case, we are truly using the grouping property of a bar graph to respond to a prediction about the relationship between fat content and moisture, as shown in Figure 3.5, "The Effect of Fat on Cookie Weight, as a Bar Graph."

Notice that all the default formatting aspects provided by typical graphic software have been removed from Figure 3.5, elements such as gridlines, legends, color, and filling. This greatly improves the readability of such graphs. And, yes, plain and simple is often the most effective! Sidebar 3.6 provides some additional questions that we can use to prompt students' thinking about the aesthetics of the graph.

Bar graphs can be harder for students to wrap their heads around, in that it requires students to further group, organize, and average data, while remembering to use colors and labels thoughtfully. However, once the concept is clear, it is a powerful way to present data that goes beyond the line graph and present a summative assessment of the data.

Sidebar 3.6 Discussion Surrounding Categories in Bar Graphs

A great discussion to have with students about a bar graph is one surrounding categories that we have to create when making bar graphs, which is where most of the important decisions have to happen. There are two sets of questions I ask, some about the number of categories, some about labels of categories. For number of categories, we can ask questions like:

- What makes a good category for a bar graph?
- Are two data points enough for each category?
- What happens if we make too many categories?

> If each data point is its own category, or too many categories are created (little fat, medium fat, high fat), we get pretty close to a line graph, and line graphs might become more meaningful because we are looking at the data in a continuous way. If too few categories, it might be a bit misleading.
>
> For the labelling of categories, we can ask questions like:
>
> - What are appropriate labels?
> - How do labels potentially influence the reader?
>
> In our example here, is 15.8% fat content really "high fat"? In such cases, the labelling of the categories becomes essential to avoid misleading readers.

Considering the Benefits and Constraints of Various Graph Types

Table 3.3 summarizes the advantages and disadvantages of each graphical representation presented above, for a quick reminder of what was presented in this section, along with the key teaching points that I typically emphasize in my classroom. For more tips about the ways to help students gather and represent their data in various ways, please see the additional links that we have provided on this chapter's page in the companion website.

Table 3.3 Summary of the Advantages and Disadvantages of Each Type of Data Representation

	Advantages	Disadvantages	Teaching Points
Table	• Very quick to create • Allows for a large amount of data in a small space	• Hard to interpret • Visually unappealing for readers	• Good for storing information, not for revealing patterns
Line graph	• Reveal patterns quickly • Quick to create	• Cannot put multiple variables on one graph easily • Can be misleading if scaling is off	• Meant for continuous data, not categories • Watch out for proper scaling of axes • Watch out for unnecessary design elements
Pie graph	• Intuitive to read • Look visually appealing quickly	• Most data sets do not suit themselves to this data representation	• Meant for a very specific type of data to show subsets of populations
Bar graph	• Powerful data representation due to many options • Reveal patterns quickly • Allow for grouping of data for further comparison	• Require some decisions regarding categories • Require further handling of data when grouping • Not meant to display every single data point	• Need a conversation regarding meaningful categories • Watch out for unnecessary design elements

How to Assess Infographics

As with all kinds of assignments and assessments, what we value at any particular moment depends on a number of factors. If asking students to create initial representations of data in their notebooks during one class session, we assess on different factors as compared to a final, fully formed graphic that is created using software and built over a series of days or weeks.

For the more formative assessments of graphs, we generally consider:

- Whether the proper data representation (bar, chart, pie, etc.) was used;
- Whether the proper data items were represented in the graph; and
- The overall visual appeal of the graph (fonts, colors, scale, etc.) that lead to ease of reading.

To see how this plays out in students' work, Wiline pulled out a few different examples of graphs generated by students. The first example consists of a scatterplot from a report written by an upper level undergraduate student, before and after revisions. On the top graph of Figure 3.6, the student used a scatterplot to represent the data about snowshoe hare behavior in the rough draft. While they used the correct type of graph (a scatterplot), the graph was hard to read for several reasons. First, the data points were linked, which made the graph really busy. Second, the axes labels were vague and a bit misrepresentative. Fixing these two elements created a cleaner graph, as can be seen in the bottom panel of Figure 3.6.1 and 3.6.2, with a much clearer "story to tell." As for any work, students can really benefit from receiving feedback on such assessments.

Figure 3.7 shows a bar graph generated by a high school student, Jessica Davis, to summarize the field experiment she had run on chipmunks. Jessica examined whether chipmunks responded differently to their different predators based on the amount of food reward they were presented with in the field. Many decisions went into such a bar graph: the student had to decide whether to lump her data based on the treatments (which predator stimuli the chipmunks experienced while foraging), or based on the reward the chipmunk was experiencing (low amount of food vs. high amount of food). She chose to design her graph to highlight the difference between her treatments, and separated the food reward by color. Each bar represents the time spent by chipmunks at a food patch, averaged across multiple chipmunks. The reasoning that Jessica used to create this bar graph, along with the logical steps taken to achieve it, is something that simply could not be replicated by any type of worksheet.

72 ◆ Visual Explanations with Infographics

Figure 3.6.1 and 3.6.2 Scatterplots Generated by an Upper Level Undergraduate Student (top panel: before revisions; bottom panel: after revisions)

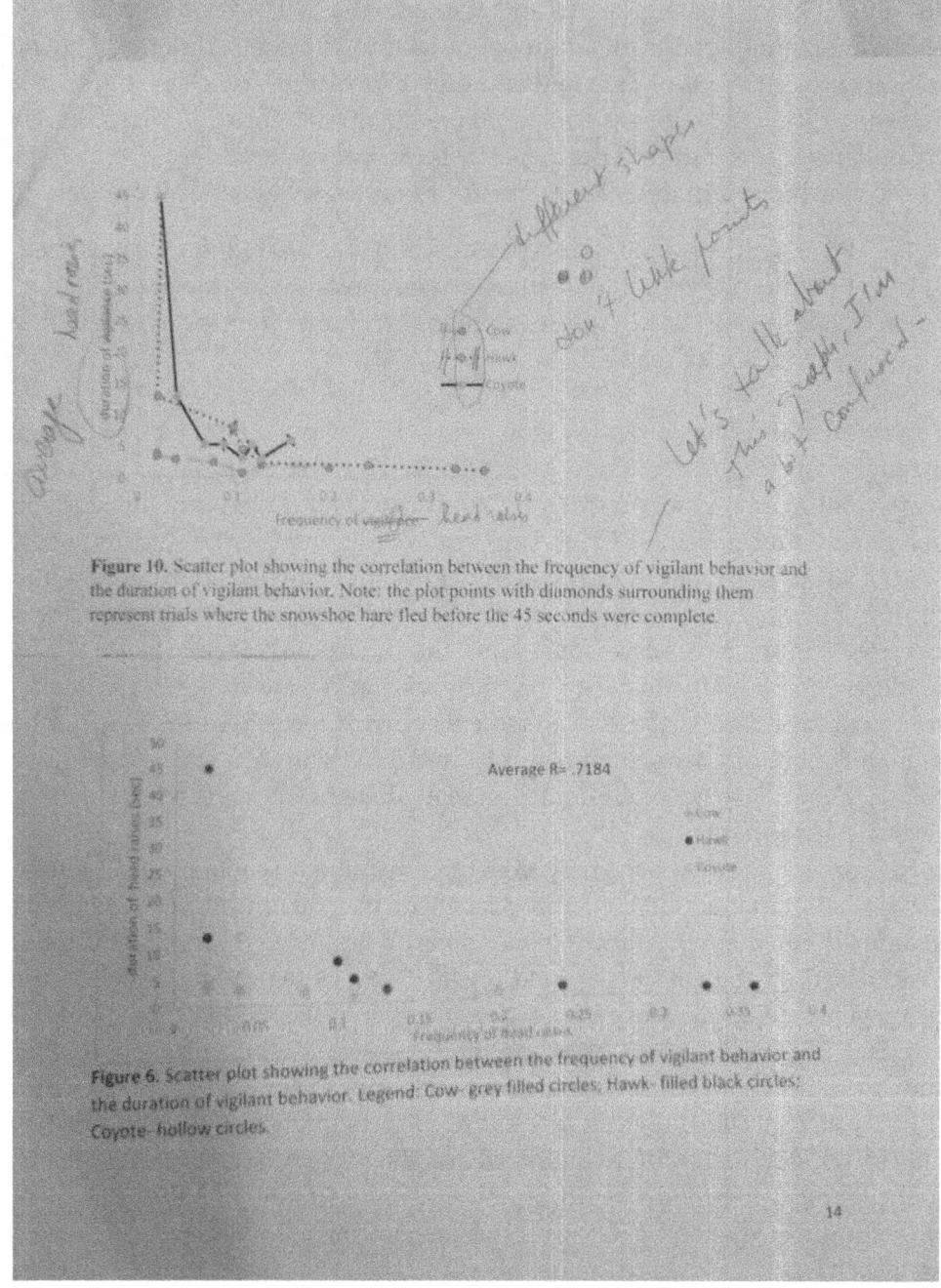

Figure 3.7 Bar Graphs Generated by a High School Student

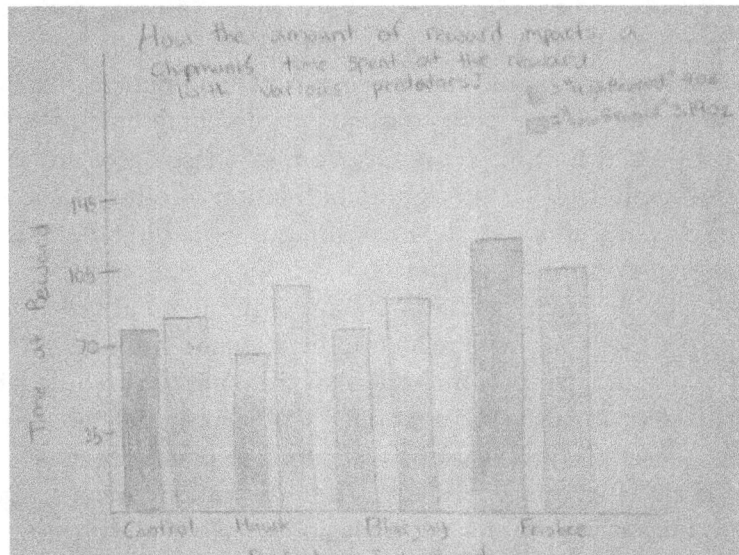

For more summative assessments of graphs, where students have had time and space to develop their ideas through the use of software, we may also consider additional elements. They can write about:

- Whether the patterns described by the students are correct interpretation of the data displayed.
- How appealing the graphs are to the reader.
- The "story" that the data are trying to tell.
- Other forms of media, especially the use of images and clipart.
- Whether the default options, often obtrusive to readers, were changed.

In the end, we boil the set of questions for our students down to this: *with the information that you have gathered, how can it best be displayed?* For a group of additional resources that relate to the assessment of infographics, curated from other educators, please visit this chapter's companion page is available on the website <https://jeremyhyler40.com/science-and-literacy/chapter-3/>.

Next Steps with Infographics

In addition to — or perhaps even before they begin to — designing their own graphs and infographics, we can also invite students to examine existing graphs. As noted earlier in this chapter, we document a strategy that Troy has adapted from one of his colleagues for analyzing and reflecting on the information in a graph, and we also reiterate how useful Kelly Turner's "Graph of the Week" can be in this regard, too.

Given that students are increasingly exposed to numerical data in relation to everything from political polls to trends in climate to the quality of foods that they eat — and that the representations of these data can be skewed, either in the graph itself or through manipulation of other visual elements such as scales, colors and fonts, text and images — helping them design and interpret their own graphs is crucial.

Returning to the ISTE standards, we are reminded that technology tools can enhance the goals we have related to literacy and numeracy. In particular, ISTE has a standard for computational thinking that requires students to "collect data or identify relevant data sets, use digital tools to analyze them, and represent data in various ways to facilitate problem-solving and decision-making" (5b) and for creative communication which suggests that they "communicate complex ideas clearly and effectively by creating or using a variety of digital objects such as visualizations, models or simulations" (6c). With both of these standards, infographics can help meet the goal.

In particular, a number of infographic design tools are available, many for free with limited (yet still very useful) functionality. Once again, for direct links to all these resources visit the companion page for this chapter. Some popular and useful tools, in alphabetical order, include Animaker, Easelly, Infogram, Piktochart, and Visme. Additionally, having students view excellent examples of infographics as mentor texts is a useful strategy. While there are countless resources available, we are particularly fond of the examples shared by Nathan Yau on his site, *Flowing Data* as well as David McCandless and his team at *Information is Beautiful*, both linked from the companion page. As always, there are many more places to find beautiful and engaging infographics, often just a web search away.

In looking at other representations of data — and by creating their own through notebooks and using technology tools — students can begin thinking more and more about the opportunities that data can provide for making more substantive scientific arguments. In doing so, they can also collaborate with their peers to discover the best ways for engaging with that data, a process that we explore in the next chapter on modeling instruction with whiteboards.

4

Encouraging Collaboration through Whiteboard Modeling

A productive hum — maybe even something a bit more like a buzz — can be heard as soon as you walk into the room, and students are gathered in groups, leaning over tables, pointing excitedly and all focused on the task at hand. Being in the science classroom, you wonder what they might be looking at: a frog dissection? Something creeping and crawling under a microscope? A chemical reaction exploding in a beaker?

As you get closer, you notice that some of the students are holding dry erase markers. There doesn't appear to be a tray, or a microscope, or any other apparatus. While there is one student who seems to be coordinating the work, each one of them is leaning in and jotting down ideas on a whiteboard, approximately two foot tall by three foot wide. All are engaged in the task at hand, representing data from yesterday's experiment into a graph, table, or … what's this … a comic strip?

Welcome to a Classroom with Students Engaged in Modeling Pedagogy

At its most basic, modeling is a technique where teachers introduce a concept and students display their learning on whiteboards, typically working in pairs or in small groups. Students then share their learning to the rest of the class for discussion (which we often refer to as "board meetings"). For Wiline and Jeremy, they often use the term "modeling" and "whiteboarding" interchangeably.

The American Modeling Teachers Association's website describes "The Modeling Method: A Synopsis." Written for parents, their clear definition states that modeling is "an approach that emphasizes the construction and application of conceptual models as a way of learning and doing science" (American Modeling Teachers Association, n.d., emphasis in original). In other words, students do not just "sit and get" content, they are learning new concepts through inquiry and showing how to apply what they have learned in the science classroom.

More specifically, modeling is a teaching method that accomplishes five goals:

- Incorporates higher order thinking;
- Generates inquiry;
- Creates student-centered learning;
- Improves discourse/collaboration; and
- Promotes risk taking.

First, teachers are constantly trying to get their students to be critical thinkers and move beyond the typical kinds of questions that are put in front of them (remember Wiline's concerns about textbooks in Chapter 1?). Modeling instruction opens the door to gently push students to develop higher order thinking skills. As students are addressing certain questions, they are asked to do much more than just give an answer. They are asked to think about their answer, often pushing them beyond just getting it correct. Students must apply what they have learned to other concepts.

Second, one of the most fantastic aspects about modeling is the inquiry that occurs with students. As students are exploring topics within the curriculum through using their science notebooks, creating whiteboards, and having board meetings, they start to ask questions about other questions pertaining to the topic. Compared to the traditional "sit and get," whiteboarding creates a new intensity at which students are asking further questions about science. For Wiline, modeling and whiteboarding has pushed students to a different level that simple lecturing has not done in the past and, as Jeremy has noted already, his experience learning from Winsor has been heavily influenced by the use of whiteboarding.

Third, though sometimes direct instruction is needed in the science classroom, neither Wiline nor Jeremy have ever been ones to have all eyes on them at the front of the class. Throughout the years of teaching science and language arts, Jeremy has always tried to at least guide the students to lead their own learning. Modeling does a phenomenal job of placing the role of teacher as a guiding force for students and requires the teacher to not just give students the answer. This is challenging to students at first because they want you as the teacher to give them the answer instead of them using higher order thinking skills, or applying what they already know to a new concept or topic.

Fourth, prior to implementing modeling instruction into science, both Jeremy and Wiline noticed that students weren't necessarily willing to step outside the box and take risks by sharing with their peers, being more creative in their science notebooks or speaking in front of their peers about the ideas they have. Modeling instruction allows students to express their ideas without scrutiny from classmates because, with modeling, students can freely share their ideas whether they are right or wrong. When they are sharing their ideas, Jeremy makes sure to write what they are saying on his whiteboard in front of the room. Yes, there may be some scientific misconceptions, but leaving those on the board until students work through them during the unit is critical. In this exploration, they realize those statements written on the board are misconceptions and, together, we erase them or rewrite them, and in so doing those statements become more accurate.

Fifth and finally, a worksheet doesn't promote discourse in any classroom, let alone a science classroom. Forced conversation that is directed by the teachers usually produces the same results; the same kids raising their hands every time a question is asked by the teacher. Furthermore, there is absolutely no conversation amongst students, only the questions are being answered. Through modeling, every student has an opportunity to discuss their thinking in a non-threatening environment. Every time students participate in creating a whiteboard, there is what we call a "board meeting." Students organize the classroom so that it is a Socratic-like discussion where they are facing each other in a circle. Each small group has a chance to discuss what they have on their whiteboard.

Considering the occasion of whiteboarding as a writing opportunity, we outline the MAPS of a typical modeling session in Table 4.1.

Table 4.1 The MAPS for Writing Through Modeling

Mode	A collaborative document, what students represent on the whiteboard often includes both numerical and visual representation of ideas, as well as words to describe and explain.
Media	A whiteboard with dry erase markers, though chart paper or drawing in a digital program like Google Draw or OneNote could be an option, too.
Audience	As a means for collaboration, questioning, and critical thinking, whiteboards are first and foremost for the participants in the group, as well as to be shared with the rest of the class during a board meeting.
Purpose	To analyze, synthesize, and represent data in graphs, tables, or other visual forms. To describe, predict, and reflect upon a scientific phenomenon in text.
Situation	Usually conducted in 5–15 minute sessions, followed by a board meeting, whiteboarding can be implemented at various stages of an inquiry to help students clarify and articulate their thinking.

78 ◆ Encouraging Collaboration

Many teachers have bought into this instructional strategy because it is no longer about the teacher standing in front of the classroom lecturing to students. According to Jackson, Dukerich, and Hestenes (2008):

> A key component of this approach is that it moves the teacher from the role of authority figure who provides the knowledge to that of a coach/facilitator who helps the students construct their own understanding. Since students systematically misunderstand most of what we tell them (due to the fact that what they hear is filtered through their existing mental structures), the emphasis is placed on student articulation of the concepts.
>
> (13)

They go on to make the point that "[s]tudents have to make sense of the experiment themselves. The instructor must be prepared to allow them to fail" (13). While this can sound a bit harsh, students show that the outcomes are generally positive, as shown in Figure 4.1 (and will be shown again below in Figure 4.3). Our hope is that the examples shown throughout this chapter show that, while an initial risk needs to be taken, the rewards of modeling and whiteboarding are great for both students and teachers.

Figure 4.1 Wiline's Freshman Biology Students During a Modeling Session (Image by Wiline Pangle)

Sidebar 4.1 Tips for Creating a Do It Yourself Whiteboard

To create the whiteboards, students will need to have whiteboard markers, erasers, paper towels and cleaning spray. In addition, students will need a whiteboard. The easiest and cheapest route is to go to your local lumber yard and buy a shower board. Then, have them cut the shower board into four equal pieces, 36x24 inches. A few more details:

- Before students use them, the shower boards will need to be waxed using something like Turtle Wax. By waxing the board, students can erase them easier and they will last much longer.
- Besides taking care of the whiteboards, teachers will want to invest in a lot of whiteboard markers. We recommend having a variety of colors at students' disposal for creating the boards.
- Finally, having Clorox wipes available for cleaning the boards will save lots of time (and erasers).

Though it seems there can be some initial costs for implementing modeling instruction, that cost is small compared to the transformation that occurs with students as they learn and develop more critical thinking skills.

How to Teach with Modeling

First, before going through what happens in the classroom with whiteboarding, let's cover some logistics: while some very nice whiteboards with cleaning cloths and special markers can be purchased, see Sidebar 4.1 for some tips about "Do It Yourself" whiteboards that have served Jeremy quite well in his own classroom.

In terms of logistics, we recommend groups from three to six students; not so few that the discussion is limited, not so many that all students cannot get their voices heard. While it helps to have students organized in small round tables, as we can see in Image 4.1, it is in no way mandatory and most classrooms might not be able to accommodate this set-up. It can be nice for students to actually get up, move around, and stand around a small whiteboard.

One of the challenges with groups that are larger than four is that not every student can or will contribute to creating the whiteboard. In middle school, it may be easier to assign roles for each group member.

What Happens with Whiteboarding: Preparing For, Conducting, and Debriefing a Board Meeting

In a typical whiteboarding session, there are three distinct phases, no matter the topic. First, students need to be given some time in small groups to brainstorm and discuss their ideas before plopping it all on a whiteboard. Then comes the class sharing time (which we refer to in this chapter as a "board meeting," a term introduced to us by our colleague Jeremy Winsor), where all students get to see the work of all small groups displayed on boards. Finally, there needs to be closure to what happened in the board meeting where students can ask questions or make comments that may not have been shared in the board meeting. This can be in the form of a whole class discussion or students could use their science notebooks to reflect on what they learned.

Before the Board Meeting: Working in Small Groups

For a successful whiteboarding session, students need to be given plenty of time to digest the question(s) in small groups. Students should have a clear idea of what is being asked (clear question), but not necessarily how to display this information; in fact, creative displays should be encouraged. Students may not always have a question to answer. For example, they may be asked to conduct a lab or experiment and display the results of the lab on their whiteboard. Furthermore, they may also have to create a graph, table, or chart based on data they have gathered from their own experiment.

During the Board Meeting: Listening Attentively

Board meetings are an integral part of the modeling process. It is during the board meeting where students not only discuss their own board that they have created, but they discuss their classmate's boards. Discussions include items they notice on other's boards, similarities, differences, misconceptions, and connections they may be noticing. Board meetings are student directed and the teacher does not lead or interrupt unless it is necessary. Even though it was originally designed as a tool for empathic listening in therapy or other conversations that could become contentious, the "LARA" method of "listen, affirm, respond, and ask" could be adapted here, too. For a brief summary of this method, a simple web search will yield numerous resources.

As noted later on in this chapter with the Board Collaboration Rubric in Figure 4.5, we want students to be engaged and ask relevant questions and make comments to each other. Asking questions such as "Why did you use that color for your chart?" is not a question that is on task, or relevant to what is being discussed. The board meeting is designed for students to make connections between what one group has done compared to others. It is a time for collaboration, an important life skill that students need for the workplace when they get older. Also, we want students to be listening to each other. Students should be listening to what others are presenting and identifying any misconceptions that may come out of their own whiteboard.

The Board Collaboration Rubric was created with the help of the science department at Fulton Schools where Jeremy works. It helps to keep students on task and helps them to understand that, with the modeling pedagogy, they will need to be active in the learning process. In addition, because there are not a lot of opportunities for students to complete worksheets or other items for grading, it gives them an opportunity to earn some credit in class that is relevant to real world skills. More on assessment is discussed later on in the chapter.

After the Board Meeting: Reflecting on Our Learning

It is important to have students reflect on what was discussed in board meetings. Earlier, we mentioned that teachers can have whole group discussions or simply have students use their science notebooks to reflect on what they learned and took place in the board meeting. One of the main things that comes out of students reflecting on their learning are questions that they think of as they have worked through the lesson. Often, it is a good idea to put students' questions on the board to either have them explore the questions on their own outside of class (or the questions may be answered later on in the unit).

Besides the questions the students may have, students may also start making connections to other lessons that have been done in class, or they make connections to things they notice in the real world. All of this leads to the point that, in the end, we are asking students to write. In this case, it is impromptu writing with classmates, yet it is writing nonetheless. Still, students sometimes have concerns, and we provide an overview of some common ones in Sidebar 4.2.

With these ideas about inquiry-based learning in mind, we now shift to Jeremy and Wiline's experiences using modeling and whiteboarding in their classrooms.

Jeremy's Perspective: Modeling in Middle School

As we described in the introduction, it's not your typical science classroom. Students are collaborating, inquiring, writing, drawing, and discovering new theories and ideas. This is my classroom on a daily basis with modeling instruction. My students are moving, thinking, observing, and discussing the science content at hand. They are even challenging each other's ideas with their own claim, evidence and reasoning about the content at hand.

I was first introduced to modeling instruction with whiteboards approximately three years ago by my colleague Jeremy Winsor. Winsor had been encouraged by another staff member to take the three-week modeling training, sponsored by American Modeling Teachers Association (AMTA), during the summer months. For Winsor, he was frustrated with going to multiple workshops on NGSS and not really being told *how* to implement them in practice. In addition, he knew students weren't doing enough application in the classroom with science concepts.

Upon completing the training, Winsor both implemented the instructional strategy into his own classroom and, moreover, he brought the idea to other teachers in our Beaver Island Institute (described in the Introduction, and in more detail in Chapter 6). It is also a pedagogy that has been adopted schoolwide for science instruction at our school. At this point, three years into my second round of science teaching, I find that I am using modeling on average twice a week.

During modeling sessions, teachers are meant to be a guide, someone to help point students in the right direction rather than giving all the information and answering all the questions. Typically, students are given a question or idea to think about. At the beginning of the year, for example, I gave my

Sidebar 4.2 How to Address Student Concerns About Modeling Instruction

Students may express hesitation, and we often see that those concerns are really manifestations of some fears they might have about performing well in class. As examples, here are a few of their typical worries:

- "I'm not good at drawing" or "Can you tell me what to draw" reflects a fear of doing something wrong when they are used to a worksheet approach.
- "I prefer not to work with a group" also demonstrates a concern that, by working in a group, they may do something incorrectly, leading to a wrong answer.
- "I don't know what data to share" again shows that students are accustomed to being spoon-fed the exact procedures and specifications for sharing data, and do not want to make mistakes.
- And, finally, some students may express frustration or even anger, noting that "You haven't taught us how to do this," which is a broader concern that they may have about moving into an inquiry-based approach for learning.

While we don't have perfect answers to all of these questions, we do want to share some of the ways that we talk to students about their concerns. We might offer:

- When worried about drawing skills, we might reply: "We're going to do this many times, it will get easier," "There are no grades on this today, there is no pressure," "The more you invest in this now, the better your grade later," or "Here's a starting point…"
- Concerned about group work, we remind them that science is a collaborative enterprise, and that they are also learning soft skills that will be useful for future employment.
- When confused about what data to share, a response like this could be helpful: "Try two different variables, see where it takes you. If you realize it doesn't make sense, you can always cross it out and start over (no one will ever know)."
- Finally, to assuage their fears about these very different methods of teaching and learning, coming up with a patent response like "I'm not giving you the answer, I am here to guide you and to help you arrive at an answer with the information that you have…"

7th graders an overarching question, "Where does the mass of plants come from?"

This approach is different than a typical science classroom, where teachers may have presented a similar kind of question, but then — over a series of lessons — provided the answers for students in the material that was presented. In an inquiry-driven approach, they are working to ask questions, collect data, analyze that data, and draw conclusions. This is a bit different from what they are used to and, sometimes, they hate it! But, this is the work of real scientists, and it is not just the common practice of reading from a book and doing a pre-fabricated lab. In many ways, this approach is right in line with what the NGSS promotes, including their focus on phenomena, cross-cutting concepts, and inquiry.

To elaborate on each of the five points introduced above about why modeling is effective, I will share a bit more about our first unit in 7th grade that deals with plant mass (as described first in Chapter 2). Students are given seeds to grow and conduct an experiment with multiple variables to test growth. While conducting the experiment they are recording data in their science notebooks (see assessment section below) and observing the plant and trying to come to a scientific conclusion of where the mass of plants come from. As mentioned before, Chapter 2 outlines the first unit I do with my students and how they effectively use their science notebooks. Here is how modeling and whiteboarding helps us meet the five goals outlined above focused on higher order thinking, inquiry, and collaboration.

As a case in point, in our very first unit this past school year students were trying to answer the unit question about what makes up the mass of plants. While they were working through the unit, students were asked smaller questions about the parts of plants and what plants used to create energy. Students studied plant leaves and wrote down what they observed and then we conducted a board session where deeper conversations could occur. Here, students could apply what they learned to the unit question which was where does the mass of a plant or tree come from? (Figure 4.2).

This year I have seen so many great questions come from my students pertaining to the topics we have been doing in our curriculum. For example:

- Can trees and plants feel pain?
- Wouldn't hunting and fishing affect population in an ecosystem?
- If hydrogen and oxygen are flammable, why is H_2O not flammable?
- Are biological and organic weathering the same?

These were just some of the great questions that have come out of our modeling instruction this year. Some have been written in notebooks, asked in class discussions, or have been brought up in board meetings.

Figure 4.2 Sample Whiteboard from the Plant Unit

My students get very frustrated when they ask me a question about something and, in return, I answer with a question (or I ask them simply, but not sarcastically, "What do you think?"). Then, I will continue to challenge them by having them discuss their question with a classmate. Eventually, students start to think for themselves and they start to take more risks, but modeling has to be done every day in order for students to get accustomed to this type of instruction.

One of the best risk-taking moments I saw this past school year was when we were dealing with the growth of radish plants, and I introduced back in Chapter 2. One group of students were using a control plant where they had the same amount of sunlight, water, and soil in a Solo cup. Then, students took another Solo Cup with radish plants and changed a variable. One particular group decided to try Powerade instead of water. They took this risk because they felt that Powerade was made up of mostly water. Furthermore, the group thought because there was some nutritional value to the Powerade that could help the plant.

It was this kind of risk taking that allowed the students to test their hypothesis that otherwise might have not been done if they were to just sit and get lectured during the 50 minute class period. In my classroom, I tell my

Sidebar 4.3 Questions and Clarification Prompts for Whiteboarding

Questions/Clarification

- I am wondering about _____ because _____.
- I heard you say _____. Can you tell me more about that?
- I am really surprised by _____. Can you explain this further or give an example?

Response to Others Ideas

- I like _____. It helped me understand _____.
- I agree with what _____ said about _____ because _____.
- To build off what _____ said, I think _____.

students that everyone needs to be ready to present their boards. One of the important guidelines I give to my students is if you can't explain it, you can't use it on your board. This helps students stay focused and on track with the content being learned. Sidebar 4.3 highlights some question and comment starters students are given to help guide them in modeling instruction.

Wiline's Perspective: Whiteboarding with Undergraduates

My experience with whiteboarding in the college science classroom is very similar to what Jeremy has already articulated. As shown in Figure 4.3, my classroom, during a whiteboarding exercise, looks, sounds and feels just as what Jeremy describes: loud, with a high level of energy, and lots of risk and inquiry happening.

There are a few things I would add about the modeling/whiteboarding technique that makes it invaluable for me. To obtain higher order thinking from students, it allows me to identify misconceptions or areas of struggle very quickly. When I walk around the room during a whiteboard session, or I listen to their arguments in a board meeting, it very quickly gives me "the temperature" of the room: what the students have clearly integrated already (thus, I can move on), what the students may need further instructions on, what misconceptions still remain that I might have to address more directly. Because it is completely driven by students, the boards display where they are in their reasoning at that point in the semester.

In addition, the students tend to be aware of their learning environment. They know that our classroom is a safe space, that they can make mistakes, that their boards don't have to all be the same. If their board happens to be incorrect, there are no negative consequences; in fact, the act of correcting the board, of rectifying what was written for all to see, is an amazing way to prompt students' learning. I often joke that the students with the incorrect boards will be the students that score highest in the matching exam question for a given unit!

Over the years, I have found that I tend to use whiteboarding for a few specific recurrent topics. First, I use whiteboards a lot for graph interpretation. On the projector, I put a graph up for the whole class to see, with axes labels but no titles, and I ask students to whiteboard the answer to the question "What is the story told by this graph?" (Figure 4.3 is a picture of my students at work). I repeat this often (in fact, during most class meetings), so much so that students get really comfortable with the question. I found this to be very effective for students to get used to graph interpretation, and it

Figure 4.3 Undergraduate Students Working Through a Graph Interpretation with Modeling (Image by Wiline Pangle)

addresses a lot of the problems I have found with visuals (which we developed in Chapter 3).

Second, I typically move from graph interpretation to graph making; it's an organic transition that can happen once students are comfortable reading graphs. Whiteboarding offers an advantage over making graphs on a piece of paper for several reasons: it forces students to collaborate, to take a stand (instead of simply waiting for me to draw the graph on the board); but it also is erasable. It can be completely wrong, then quickly fixed without anyone knowing. I further explore in Sidebar 4.4 the different times at which I use whiteboards over notebooks in my science classroom.

Third, I love to use whiteboarding for brainstorming sessions that have some higher-order thinking involved. In one typical kind of lesson, I often present a pattern from nature, then ask students to generate, on their whiteboard, a few hypotheses about why this pattern might be there. The board meeting that follows is usually rich in ideas and can easily be followed by a second whiteboarding on the type of data we would need to figure out which hypothesis is supported. As such, whiteboarding provides a great tool for cross-teaching concepts.

Sidebar 4.4 When to Use Whiteboards Over Notebooks?

Whiteboarding presents some interesting opportunities that are different from notebooking. While we covered the value of science notebooks in Chapter 2, it is interesting to contrast the approach of notebooking with that of whiteboarding.

Notebooks provide a somewhat "permanent" trace of the students' thinking at a certain point in time. On the other hand, whiteboards are extremely fleeting in nature. I have found that both have their space in my science classroom, but that I use both of them somewhat differently, which I tried to capture below:

When to whiteboard:

- When students are still learning to work with each other.
- When I want students to collaborate and reach a consensus.
- When students are still exploring a topic.
- When I have not lectured about a specific topic much.
- When I don't need a specific entry or trace of the answer, but instead am after the process.
- When I think there is a high likelihood that students will get the question wrong.

When to notebook:

- When students need to develop a personal opinion.
- When students have reached a consensus.
- When the classroom is loud, and we need to re-center the energy.
- When we have had many discussions about a topic already.
- When I'm looking for a more permanent trace of the answer for exam purposes.

Lastly, I use whiteboards extensively in reviews before exams. I find that once students are comfortable whiteboarding, it provides the perfect set up for a safe, yet productive review session. As I outlined above, it also allows me to quickly assess what students might really need from me to succeed in an exam, as it exposes misconceptions quickly. I use different types of writing prompts for students to review material on whiteboards; that is, I might ask for students' "muddiest points" from lecture materials. It's interesting for students to see the similarities between boards. I often have students work together to generate what they would consider hard, yet fair exam questions; the following board meeting turns into having students answering other groups' exam questions. Both of these examples emphasize one aspect of a review session that cannot easily be replicated without whiteboards: the collaboration that happens between students. Students are generating the muddiest points together, answering each other's board questions, and really, in the end, teaching each other the material.

One caveat worth pointing out is that whiteboarding is time consuming. I never manage to have a whiteboarding session that lasts less than 15 minutes. Considering that my lecture time in a college setting can be as short as 50 minutes, that is a large chunk of time for me. I have tried, very unsuccessfully, to shorten my whiteboarding time: it only generates annoyed students that felt rushed, and a very frustrated instructor, too. Instead, I've now embraced the fact that to happen in good conditions, whiteboarding does take time, and so be it. It is too essential a technique in my classroom to abandon, but I do choose intentionally, because of the time, when to whiteboard. In a given semester, with classes that meet twice a week during sessions lasting just 75 minutes, I attempt to whiteboard at least once a week. Sometimes it goes in spurts, with lots of whiteboarding (during each lecture session) for three class periods in a row, before we take a break from the practice due to the nature of the material we are covering.

How to Assess Modeling and Whiteboarding

When it comes to assessing students during the modeling process, typical assessments are thrown out the window. Returning to Jackson, Dukerich, & Hestenes, first introduced above, they note that:

> Whiteboarding is another major component of assessment. In whiteboarding, small groups of students write up the results from a lab or solutions to a worksheet problem on a small whiteboard. One student from the group presents the whiteboard to the class, responding to questions from the class and the instructor… this is an important assessment tool that helps the instructor determine how well students have mastered the concepts.

(2008, 14)

Put another way, students don't have a science book, nor do they complete a worksheet every day for a teacher to check over. Any sheets that students receive are designed as a way to get them thinking about a new science concept. Often, students are working with each other on what is handed to them by watching a short video presented in class, or they may visit a few reliable websites to gather some information about the topic. Students may take as long as 50 minutes to complete tasks that are involved with any worksheet that is given to them, and are then assigned a completion grade.

For Jeremy, science notebooks and the occasional worksheet aren't the only assessments that students see throughout the year. The students do take a few quizzes throughout each unit and they do have a final unit test. However, the assessments the students have are not the traditional ones that involves multiple choice or matching questions. Quiz and test questions are open-ended and students are required regularly to draw graphs, charts, pictures, and diagrams.

Below is an example quiz from the plant unit. Students are required to read information from a given chart and asked to draw a conclusion from it. Jeremy structures key vocabulary terms as an expectation of the quiz responses, encouraging students to use evidence and reasoning. The assessments, when paired with modeling, allow students to explain their ideas through writing. Yes, that is the English teacher in him, and Jeremy's students are allowed to think through an answer and dive deeper in their thought process in this kind of quiz where a multiple-choice assessment does not allow them to meet these cognitive goals. An example of this kind of quiz is shown in Figure 4.4.

Leading up to quizzes and tests, students work collaboratively to complete whiteboards to model what they have learned and what they want their classmates to learn from them. Recall that students are not graded on what they put on their whiteboards or the quality of their drawings, charts, or graphs they put on their whiteboards. When it comes time for our "board meetings," we collaborate as a class and the students present their findings on their whiteboard. This leads to an assessment of performance in the board meeting, and Jeremy utilizes his Science department's board collaboration rubric in Figure 4.5.

Wiline's Perspective

While holding middle school students accountable is a priority for assessment, the focus does shift a bit when working with undergraduates. I do not specifically grade any whiteboard sessions, or board meetings. These are simply part of my students' overall participation grade, and I do not keep track, specifically, of who contributes. However, I do write exam questions that closely match the whiteboarding activities that we have done in class.

Figure 4.4 Quiz Designed to Elicit Student Thinking after Whiteboarding

Directions: Answer each question to the best of your ability. As needed, include drawings, diagrams, and written descriptions to better communicate your thoughts.

1. During the mid-1600's, Von Helmont grew a willow tree in a pot for five years. Only water was added during the experiment. The experimenter found these results.

Year	Weight of Tree	Weight of Dried Soil
1642	5 pounds	200 pounds
1647	169.1 pounds	199.9 pounds

   Based on this evidence, as the tree grew, do you believe that most of the tree's mass came from nutrients in the soil? Explain your reasoning.
2. What is a seed? What are its three (3) basic parts?

Figure 4.5 Board Collaboration Rubric from Jeremy's Science Department

Board Collaboration Rubric

- 3/3 Points: Student provides unprompted relevant input **or** questions. Student input and behavior is appropriate and respectful. Student is an active listener and engaged in the discussion.
- 2/3 Points: Student provides relevant input **or** questions **only when prompted**. Student input and behavior is appropriate and respectful. Student is an active listener and engaged in the discussion.
- 1/3 Points: Student appears to be listening but does not provide relevant input **or** questions **even after being prompted**.
- 0/3 Points: Student input is not appropriate or respectful, or student is not engaged through discussion or listening, or student behavior is otherwise not appropriate.

One type of exam question that I always include in my exams has to do with graph interpretation, since this is a skill that I work intently to develop with my students. Students have typically worked most class sessions on their graph reading skills ("What is the story that this graph is telling us?) using whiteboards. Exam questions would then present students with a brand new graph that they have not been exposed to before, and would ask them to write down what the graph is showing them. These questions could be either multiple choice, providing five different interpretations of the graph, or open ended, with students writing down, as they would have on a whiteboard, what story the graph is depicting. When grading such open-ended exam questions, I am looking for accuracy, and the ability of the student to read the graph and connect it to lecture materials.

Other types of exam questions might be more general. In one whiteboarding session, students might be asked to "Draw five water molecules interacting. Include and label all bonds involved." An exam question could then ask: "What are the bonds involved between two interacting water molecules? Within a water molecule?" Students that were involved in the whiteboarding session are much more likely to remember the structure of a water molecule, along with the different bonds involved. As another example, a modeling session could encourage students to use a line graph to depict how the amount of DNA changes over the life cycle of a cell. Then, on the exam, they would be asked: "How does the amount of DNA change over the life cycle of a cell? Justify your answer with a graph."

Next Steps with Modeling and Whiteboarding

In any whiteboarding session, we can encourage students to do more than just report what they've found in a whole group. In a blog post for Chem Ed Exchange (2014), Erica Posthuma describes a number of adaptations that we can use for making whiteboarding sessions even more engaging. In this sense, she suggests that protocols like "four square" can be used so each of the members of the group represents an idea in one smaller square on the board, or two groups can share their solutions in a "dueling whiteboards" competition. Additionally, she provides some useful question templates for teachers and students, including ones to compare and contrast solutions across groups, to point out common misunderstandings, or to engage in a "boardwalk" before the meeting begins. Along the same line, Sam McKagan and Daryl McPadden (2017) offer a number of suggestions for using whiteboarding in the physics classroom, including some diagrams to demonstrate the grouping structures they discuss.

As noted throughout the chapter, whiteboarding for Wiline and Jeremy — as well as many other teachers engaged in the practice — relies on analog tools as a way to foster collaboration. For good reasons, we can see value in having students physically gather around a shared whiteboard, use dry erase markers to draft their ideas, and then share those ideas with classmates in a board meeting. Given the socio-emotional bonds that are made in this process as well as opportunities for writing and representing thoughts visually, we hesitate to add useless technology twists. Liz Kolb, in her book, *Learning First, Technology Second*, is adamant about keeping the priorities straight. She argues that:

> Technology integration is more complex than simply using a technology tool; pedagogical and instructional strategies around the tool are essential for successful learning outcomes.

(10)

Thus, we offer a few extensions and ideas for whiteboarding, noting our preference to ground the work in the physical act of collaborating, face-to-face, in real time. Still, there are ways to share what has been created in technology-enhanced ways and to create whiteboards with additional technology tools. In doing so, we are reminded of ISTE's student standards for global collaboration, encouraging students to "use collaborative technologies to work with others, including peers, experts or community members, to examine issues and problems from multiple viewpoints" (7b) as well as to be creative communicators who "publish or present content that customizes the message and medium for their intended audiences" (6d).

We begin with a few tools that can be used to support modeling activities. There are a variety of online whiteboarding websites, as well as programs and apps like Microsoft's OneNote that can be used. Some of the options available to collaborate online for free (often with limited functionality) include, in alphabetic order: Aww, Conceptboard, Crayon, DrawChat, WebWhiteboard, Witeboard, and WhiteboardFox (links on the companion website: < https://jeremyhyler40.com/science-and-literacy/chapter-4/>). Of course, similar types of tools can be used in Google Draw, or even with the drawing tools in other office suite tools, like Google Slides or PowerPoint. As with other resources, these are linked on this chapter's companion page. Inviting students to use these alternatives, we again encourage readers to think about the pedagogical purpose — having students engage in substantive dialogue about their data and interpretations — and not to get lost in the features and functionality of apps themselves.

As another option, there are ways to bring a digital twist to analog whiteboards simply by having students snap a picture and then share it with others, opening up possibilities for annotation. As they prepare for a board meeting, one student from each group could take a picture of the group's board, then post that picture in a shared Google Slides presentation. Each group's board would be on one slide, and then classmates could use the drawing and commenting features, as well as the textbox below the slide, to provide additional comments, questions, and suggestions to the presenting group. These opportunities for annotation would probably be best after the class's board meeting, which would keep students from being distracted by their devices while trying to listen to their peers.

Of course, we end where we began, with the power of the vignette we shared at the top of this chapter. While we do want to encourage students to use technology in critical and creative ways, we also want to be mindful of the fact that the rich, dialogic opportunities that open up when students are huddled around a whiteboard can be transformative. Jeremy and Wiline have seen this in their classrooms, and we have all seen it when leading professional development with our colleagues. Reminded of this power of collaboration, we now turn our attention to some additional strategies that we can use to integrate writing into our science classrooms, bringing some opportunities for using creative nonfiction, vocabulary instruction, and even a few more apps, websites, and programs that we have not mentioned earlier in the book.

5

Additional Strategies to Encourage Inquiry, Reading, and Writing

As we open this chapter, it is worth reiterating some of the main points from Appendix M, "Connections to the Common Core State Standards for Literacy in Science and Technical Subjects" (2013b) as a way to think about the additional strategies and technologies described here. This document both outlines the many number of overarching practices that are literacy-based and makes explicit connections between the NGSS and CCSS Literacy Standards. Those practices include:

- Asking questions and defining problems;
- Planning and carrying out investigations;
- Analyzing and interpreting data;
- Constructing explanations and designing solutions;
- Engaging in argument from evidence; and
- Obtaining, evaluating, and communicating information.

One of the challenges evident in the ways that the NGSS presents writing to its readers is that, according to Maria Gigante, "the NGSS's characterization of the linkage between science education and English-language arts [is framed] as 'risky'" (2016, pp. 91–2). By framing this relationship in tension, Gigante, an assistant professor of English at Western Michigan University, believes that writing may not be seen as critical to the process of learning and doing science as it could be. She asks: "How can writing be positioned as integral to science education rather than as a risky importation that could disrupt science teaching and learning?" (95), and we are reminded that we need to keep our focus clear when bringing new literacy strategies, especially those involving more technology, into the science classroom.

In this chapter, we outline a few of our other favorite strategies that we use in our attempts to combine science and literacy instruction. Each of these activities can be woven into many of the lessons we've already presented earlier in the book including the use of notebooks, creating visuals, and engaging in the modeling process. Writing, in its many forms, can be integrated before, during, and after class, in quick bursts or longer durations. Our goal here is not to be exhaustive, but instead to show the many ways that integrating active literacy practices into inquiry-based learning can happen.

Moving into Deeper Inquiry with the Question Formulation Technique

Whenever we ask students (or adults, for that matter) to brainstorm, there are sometimes blank stares, sometimes groans, and, even worse, sometimes people who will leave the room. While this isn't common — and not likely with our students, who feel a bit of compulsion to stay in class even when they are annoyed — the simple fact is that, when it comes to inquiry-based learning, not all students will be excited, and not all ideas will then lead to substantive inquiry questions.

This is a challenge that the Right Question Institute (RQI) has been dealing with for years in order to push people, young and old, to deepen their thinking and inquiry. To that end, they have created the Question Formulation Technique (QFT), what they describe as a way to help people "formulate, work with, and use their own questions" (Right Question Institute, 2019). The process for moving through the QFT — as well as many additional resources that can be used to support that process — are available for free, once an account is created, at their website (link from this chapter's companion page: < https://jeremyhyler40.com/science-and-literacy/chapter-5/>).

Put another way, the QFT requires the teacher or facilitator to develop a question focus, which can "be a statement, phrase, image, video, aural aid, math problem, equation, or anything else that gets the questions flowing" (Right Question Institute, 2019). The point here, however, is not just about asking a question. Instead, it is work for the participants to do in the process of asking. They outline four rules for participants to follow, such as "Ask as many questions as you can," and "Change any statement into a question."

The authors at the Right Question Institute then move through a process of improving the questions, identifying whether the questions are open-ended or closed-ended, discussing the value of each question, and finally prioritizing the questions. As Nicole Bolduc, a 7th-grade science teacher, concludes in a blog post for the Teaching Channel, using the QFT over time was valuable for her students, and "students understood how important it is to develop and practice questioning skills in and out of the classroom because they realized that questioning leads to learning" (2019).

The QFT process, in summary, helps refine inquiry.

Having only learned about this strategy in the summer of 2019, Troy used it as an opening activity during one morning of our 2019 Beaver Island Institute. At the time, we were exploring the connections between macroinvertebrates and stream health, broadly considering the importance of biodiversity. Our colleague, Jeremy Winsor, provided the opening statement about the role of biodiversity in a healthy ecosystem and, from there, we jumped into the QFT. Teams of three moved through the process, working to generate, refine, and prioritize their questions. Using whiteboards, teachers moved through the entire process in about ten minutes, as shown in this image of Wiline (left; Figure 5.1) working with two of our 2019 participants, Amy Lesperance, a teacher at Shelby Junior High School in Shelby, MI (middle), and Bridget Baumgarten (right), a teacher at Dean A. Naldrett Elementary School in New Baltimore, MI.

Figure 5.1 Wiline, Amy, and Bridget Engaged in the QFT Process (Image by Troy Hicks)

From this brief activity, a sampling of questions that were generated from the groups in the first few minutes include the ones shown in Sidebar 5.1.

Sidebar 5.1 Sample of Teachers' Questions About Biodiversity Generated from the QFT

- What is biodiversity?
- What does biodiversity mean?
- Does climate affect biodiversity?
- Where is biodiversity found?
- Does biodiversity have an effect on humans?
- Can humans affect biodiversity?
- What can biodiversity tell us about the environment?
- Do seasons affect biodiversity?
- How do we quantify biodiversity?
- Why does biodiversity really matter?
- Does biodiversity look the same in every ecosystem?
- How do we measure biodiversity?
- Does biodiversity change over time?
- What are some of the factors that affect biodiversity?
- Can you change an ecosystem's biodiversity?
- How many animals does it take to be biodiverse?
- What does biodiversity look like?
- What is needed for biodiversity?
- What are the advantages of biodiversity?
- What are the factors that can harm biodiversity?
- Is biodiversity a good thing?
- In what context is biodiversity a bad thing?
- How many species in an area constitute biodiversity?
- How big of an area do you pull from to measure biodiversity?
- Do we measure biodiversity with plants? Animals? Both?
- How does biodiversity relate to me?

The QFT then wrapped up with us sharing our prioritized lists of questions. As teachers reflected on the entire institute experience, we as facilitators found it compelling that many of them reiterated the idea that the QFT was one of the most powerful activities we did. Most described the ways in which it helped them think about inquiry, broadly, as well as how to help their own students move into deeper thinking about scientific concepts and, as the NGSS encourages us to do, explore phenomena. Again, having only been introduced to the technique this summer, as we finish this manuscript we are all thinking about ways to use it in our own teaching, and this initial experience was quite promising.

One other interesting set of tools for brainstorming can be found on the website (as well as in the book), Gamestorming. Described as a "toolkit for innovators, rule-breakers, and changemakers," the Gamestorming website offers a variety of game ideas that can inspire thinking, support team-building and decision-making, and just generally make the brainstorming process more fun. While we have not tried these strategies yet with our own students, Troy was introduced to one particular strategy during a professional development event, "6-8-5," which is designed to move beyond under-developed ideas and "to generate lots of ideas in a short period of time" (Gray, 2011). While many of the strategies seem to have an entrepreneurial, Silicon-Valley mindset and tone to them, like the QFT, these strategies could be adapted for science inquiry.

Creating (Creative) Nonfiction

Many of us struggle with long lists of facts and arcane vocabulary in typical scientific writing. As an alternative, we can invite students to pursue creative nonfiction. Described by the founder and editor of *Creative Nonfiction* magazine Lee Gutkind, the goal of the genre in simple terms is "to make nonfiction stories read like fiction so that your readers are as enthralled by fact as they are by fantasy. But the stories are true" (2012, p. 6). In this sense, writers have more opportunities for using literary elements to enhance the overall effect of a story, especially in describing characters and elaborating on the unique situations and contexts in which they find themselves. The work must still be based on fact and personal experience, combined. As Gutkind elaborates, creativity is not about hyperbole; instead, he argues, "[it] is possible to be honest and straightforward and brilliant and creative at the same time" (6). Creative nonfiction, in this sense, gives our students license to explore science while also playing around with various opportunities for expression.

In particular, Troy has enjoyed reading the author and humorist, Bill Bryson, and employing some of his writing about nature as mentor texts for creative nonfiction. One popular passage from Bryson's book *A Walk in the Woods* (2006), in which he describes the natural history of black bears as he also contemplates a hike on the Appalachian Trail, works on multiple levels as it shares accurate, scientific information in a witty format. For instance, after describing how the likelihood of a black bear attack on the trail would, ultimately, be very unlikely, Bryson muses that:

> If I were to be pawed and chewed—and this seemed to me entirely possible, the more I read—it would be by a black bear, *Ursus americanus*. There are at least 500,000 black bears in North America, possibly as many as 700,000. They are notably common in the hills along the Appalachian Trail (indeed, they often use the trail, for convenience), and their numbers are growing…

> ... Black bears rarely attack. But here's the thing. Sometimes they do. All bears are agile, cunning, and immensely strong, and they are always hungry. If they want to kill you and eat you, they can, and pretty much whenever they want. That doesn't happen often, but—and here is the absolutely salient point—once would be enough.
>
> (16)

Bryson's deadpan delivery accentuates the point that bears, although they are a rare sighting on the trail, are quite dangerous, and worth a hiker's serious consideration. In this book, and many others that he has written, Bryson is able to take scientific concepts, many of them complex, and make them quite accessible, a skill that we would want our own students to emulate. His quick asides (e.g., "this seemed to me entirely possible," and "here is the absolutely salient point") reiterate the key facts that he has presented with a touch of intellect and humor, driving home the importance of those details while not completely boring or berating the reader.

Given this introduction to creative nonfiction, and Bryson's writing, Troy structures a lesson around the task of looking at a typical nonfiction passage and, in turn, Bryson's writing about black bears. Using both of these as mentor texts, he invites students to consider a few questions, as shown in Sidebar 5.2.

After exploring these questions, the students have a chance to write their own creative nonfiction. Blogger Dave Hood provides a great list of ideas

Sidebar 5.2 Questions for Comparing and Contrasting Typical and Creative Nonfiction

- With the typical passage:
    - Look for the "typical" nonfiction text features in the article. What do you see and notice?
    - Then, look for the big scientific concepts as well as the specific facts about the animal, plant, or phenomena in the article. Think about how the information is introduced to the reader. How are sentences phrased? How many facts are presented in each sentence? Each paragraph?
    - Finally, what is the most memorable phrase or sentence from the article? What makes it memorable?
- With Bryson:
    - Circle, underline, or highlight the "facts." Get a sense of how many facts are in there and see where — and how — Bryson uses them.
    - Then, consider many elements of creative nonfiction such as extensive research and obscure facts, as well as the use of narration and literary elements. In particular, try to find moments where:
        - Bryson positions himself — or you, as the reader — as a character in the passage.
        - What is the "real life" aspect of Bryson's work? What is the underlying purpose?
        - What literary elements does Bryson use? As a reader, do you feel they are effective?

about creative nonfiction elements on his site, *Find Your Creative Muse* (2010), noting that the writer should "[u]se literary devices to tell the story" and "[e]nd the creative nonfiction piece with a final, important point." Writers engage with the facts as a way to propel the story forward. For a version of this lesson as a Google Doc, please visit this chapter's companion page. Below, we provide two examples of some creative nonfiction that teachers involved in our Beaver Island Institute have created.

Creative Nonfiction Example 1: Narrative Poetry

One example of this kind of creative nonfiction writing comes from a teacher who joined us during the 2017 institute, Lexi Jensen, who at the time taught at Mosaic Middle School in Sheboygan, Wisconsin. Having learned about an invasive species, the round goby, during one of our experiments on the water in Lake Michigan, Lexi became interested in the history of the species and its effects on the lake, especially after the devastating effects of the zebra mussel invasion in prior years. Both species, having found their way into Lake Michigan, also found their way into her creative nonfiction, which begins with this:

> Humans often fear the arrival of aliens who would lay claim to everything they find. Bringing their strange weapons, rulers, and skills, the theory is that humanity would have no chance at survival once the dreaded aliens land. Well, aliens have invaded the Great Lakes, [and created] a world of their own.

She goes on to have the gobies and mussels interact through a narrative poem, trading stanzas with one another and describing their conquest, "We are triumphant. | This Great Lake is all our own. | We have all the power." Yet, by the end, she makes a call to her readers, encouraging them to fight invasive species by saying "There is no more time to wait. | We all gather. Act." As a multi-voiced narrative, told from the perspective of humans as well as the two invasive species, it works to show the complicated nature of the freshwater ecosystem and the effects all three have upon it. To view the entire story, please visit this chapter's companion page on the book's website.

Creative Nonfiction Example 2: Real Estate Ads

Two other great resources from which to draw upon are the work of writing expert Barry Lane, particularly his *51 Wacky We-Search Reports: Face the Facts With Fun* (2003), and college composition instructor Traci Gardner's

Designing Writing Assignments (2008). From these two texts, we see a number of ways in which writers might approach different genres, audiences, and purposes. In one humorous example, Lane offers a number of alternative stances, aimed primarily at upper elementary through high school grades. From "wacky wanted posters" and "trading cards" through "movie previews" and "infomercials," Lane gives students many options for creative nonfiction. Similarly, Gardner provides, especially in Chapter 4 of her book, "Defining New Tasks for Standard Writing Activities," lists of alternative audiences, positions, timeframes, sources, and genres, so that chapter is especially worth downloading. Lane provides a number of activities from the book for free on his website, and Gardner's book is available as a digitized book by the Colorado State University Libraries (links available on our companion site).

As an example, Shannon Oswald, a middle school teacher from Montcalm, Michigan, created these two "real estate ads" with us during our 2016 institute. During a day that we visited the beach and considered the biodiversity of a dune ecosystem (from the lakeshore to the treeline), Shannon took up one of the writing prompts and used this unique genre to creatively describe what she found in two areas. For the plants closer to the shore, she notes that "if erosion doesn't wear you down and you don't care for a lot of foreigners, beachfront is the place for you!" Blending humor with ecological content knowledge, she is able to make the point that hearty plants are the only ones that can survive in an exposed environment. Then, for the intermediate space between the shore and the treeline, she created an ad for what sounds like a perfect suburban neighborhood, "Valley Values." The ad goes on:

> Love meeting new people? Enjoy the easier life free from erosion and the crowding of the large trees and plants by choosing Valley Values. Located between the beach and the tree line, our community boasts the largest variety of species around. As many as 13 varieties live in just one community!

Again, integrating the idea of species diversity with a bit of humor, Shannon's work captures her emerging understanding of this concept as it plays out in the different areas of the dune ecosystem. Moreover, she was able to integrate digital pictures from the field experience into this alternative genre, moving beyond the basic summary of her experience and into a description that would likely help her better remember the concepts discussed and the data collected (Figures 5.2 and 5.3). For a higher resolution image of each, visit this chapter's page on the companion website.

Figures 5.2 and 5.3 Real Estate Ads that Demonstrate Understanding of Ecological Succession

BEACH FRONT
WE REALLY DON'T LIKE YOU IF YOU'RE NOT LIKE US

FEW AND PROUD

If erosion doesn't wear you down and you don't care for a lot of foreigners, beachfront is the place for you!

Boasting an average of fewer than 6 species per 4m square*, our waterfront properties have the exclusive conditions you're looking for in a place to put down roots. Not everyone can handle our rough conditions. If you're like us, you're safe here! *Between 4 and 7 species per unit.

XENOPHOBICBEACHPROPERTIES.CO
CONTACT JEFF AT 1 800-I-HATE-ALL

XENOPHOBIC
PROPERTIES

VALLEY VALUES
LIVE IN THE MIDDLE AND ENJOY AN EASY LIFE WITH A FRIENDLY COMMUNITY

THE PROPERTY

Love meeting new people? Enjoy the easier life free from erosion and the crowding of the large trees and plants by choosing Valley Values.

Located between the beach and the tree line, our community boasts the largest variety of species around. As many as 13 varieties live in just one community!

LIGHTHOUSEPROPERTIES.CO
CONTACT MARK AT 398 2938 398

LIGHTHOUSE
PROPERTIES

In summary, creative nonfiction provides writers with opportunities to make meaning from their observations as well as more formal data collection that moves beyond the traditional summary or lab report. Also, if provided as a formative assessment opportunity, these creative nonfiction pieces can reveal misconceptions, allowing us as teachers to adjust our instruction. Indeed, it requires the kind of synthesis and creativity that we should aspire to move students toward as we more thoroughly integrate the NGSS, CCSS, and ISTE standards.

Jeremy's Perspective: Everyday Vocabulary and Science Vocabulary

Using strategies that are mentioned above for formative assessment can help teachers see how students may not understand science vocabulary terms. Ideas such as the ones listed below are not magic bullets to fix vocabulary issues, but can help relieve the frustration that teachers often face with students not using specific vocabulary or students just memorizing it and never thinking about the words again. Teachers can:

- Help students understand that there are words that will be used in science, and there are many, many more that will be used in everyday vocabulary and writing. I (Jeremy) help students distinguish between the two types by labeling them classroom vocabulary and everyday use vocabulary. Classroom vocabulary are words that in general will only be used in a science classroom or any other subject area classroom. For example, the term "photosynthesis" will probably not be heard in a group conversation among middle school students in the hallway or outside of school! Everyday use vocabulary are words they will not only use in the classroom, but outside the classroom as well. Words such as variable or control will be more likely to come up in daily conversation.
    - By helping students determine which terms are strictly science terms, they can isolate those terms and know they will only use those scientific words in science class. This can help students also learn life skills such as how to talk in different social circles based on the words they have learned.
- Have students look up words they don't understand in articles or other readings they complete and discuss in class.
    - Before students engage in deep reading, have students scan the reading and write words in their science notebooks they don't understand and they can circle it in the given reading. Then, as a whole class, the words and definitions can be listed on the whiteboard or a shared Google Doc can be created that students can

have access to throughout the school year. The teacher could have students copy terms in their science notebooks as well.
- After the definitions of the unknown terms are discussed and students have access to them, they can then read the article with a better understanding of the vocabulary used in the reading, thus making it easier for the student to read and comprehend. By incorporating this strategy, I feel students don't focus on unknown words inhibiting them from remembering what the article is trying to teach them.

◆ Show students how to apply the vocabulary they have learned effectively in their writing (especially by using tools available in Word and Google Docs to highlight, circle, underline).
- One strategy to use with students so they just don't memorize the words for a quiz or test and forget them, is to have them use the words in short writing pieces. One example is having students write a comparative analysis explaining two graphs. In that piece of writing, it requires students to use five scientific vocabulary words such as control, variable, experimental design, photosynthesis, and stomata.
- The students highlight or underline the words so the teacher can easily spot the word and check for understanding to see if the word was used correctly in the context of their writing. By applying the words to their own writing, students are less likely to just forget the word and have a new word to add to their scientific vocabulary. Furthermore, students are more likely to identify the meaning of the words they use in their writing when they are spotted on tests, quizzes, and in future readings they may have.

Overall, these strategies can help vocabulary retention with students and essentially make it "stick" more. Also, students can use these strategies not just in a science classroom setting, but also use them in their other content areas as well.

Wiline's Perspective: Concept Mapping

Having used concept mapping extensively in my undergraduate classes as a type of formal assessment (usually as assigned homework), I find that it allows students to make a list of vocabulary on their own — they can connect concepts in whichever ways their brain functions, and gain ownership over a long list of terms. As such, I reserve concept mapping for sections of the semester heavy in vocab, such as protein synthesis or energy usage. For a given topic, I may get very different looking concept maps, all

106 ◆ Additional Strategies

in different formats, connecting things in different ways. I grade for correct links and logic, not for students to link terms in all the ways that terms can be linked.

Below are two examples of concepts maps for the same unit, protein synthesis. After covering the details of protein synthesis in the lecture, students often feel overwhelmed by the amount of vocabulary. They are asked to complete a concept map using all the vocab covered in the lecture as homework. Students are given instructions on what a concept map is, and that they should pay specific attention to the linking terms they are using to link concepts. You can see how in Figure 5.4, Kendra P.'s concept map on protein synthesis, follows an organic flow of the materials, with some terms emerging as central to her thinking while others remaining more peripheral.

Figure 5.4 Kendra P.'s Concept Map on Protein Synthesis

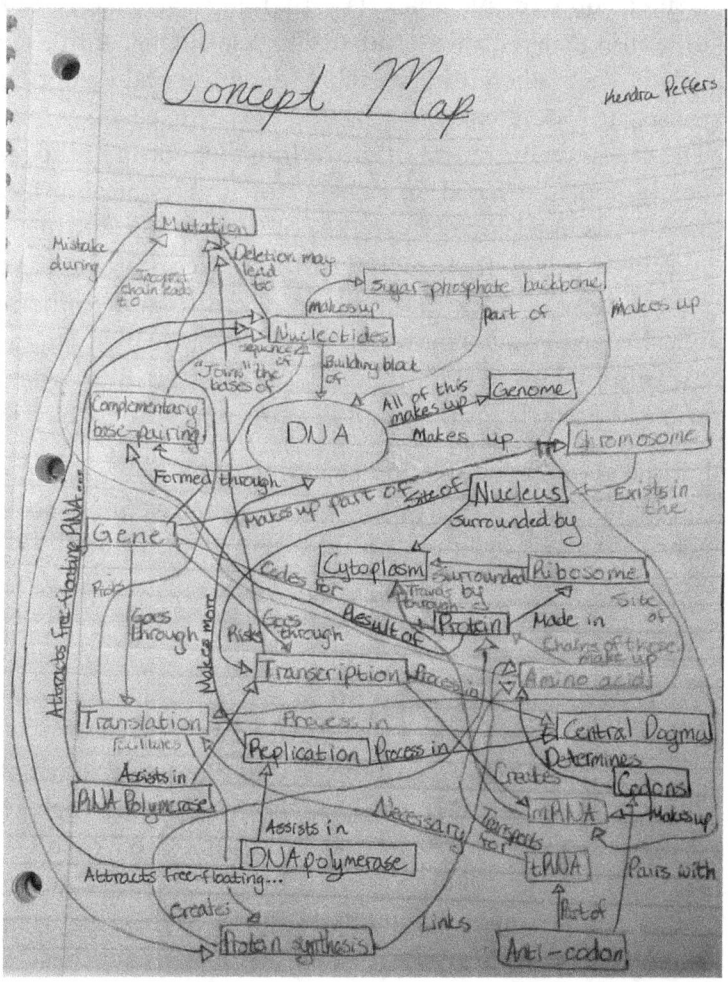

Figure 5.5 Julia A.'s Concept Map on Protein Synthesis

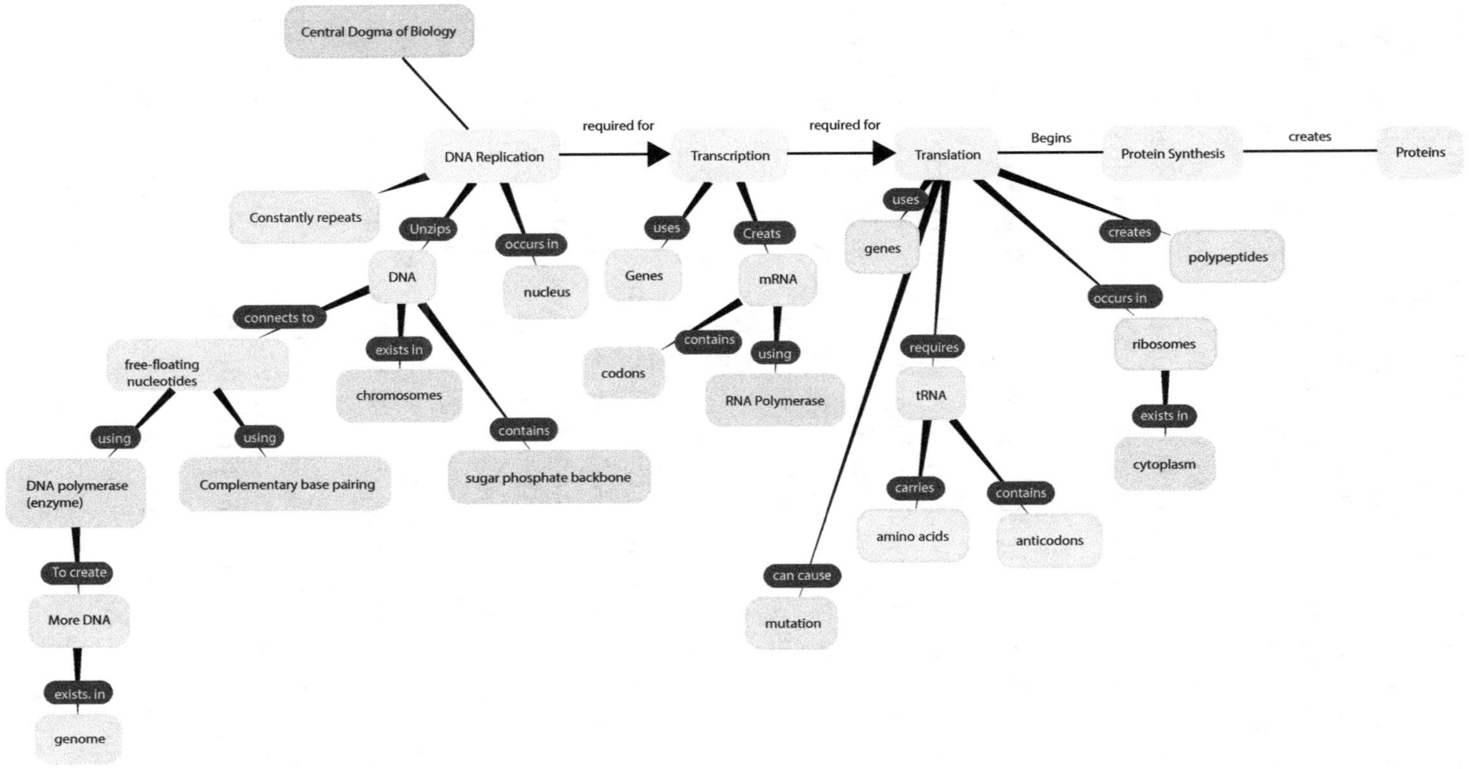

Most terms on her map are linked to many other terms, reflecting the student's thinking about these processes. In contrast, in Figure 5.5, Julia A.'s concept map on protein synthesis, you can see that the student processed the information in a much more linear fashion, organizing terms in specific categories that she even color coded (the yellow boxes). Boxes on Julia's maps are linked to only one or two other terms, highlighting specific features (especially locations, which are indicated below each process: nucleus, genes, cytoplasm).

Both concept maps received full credit for the assignment, yet are vastly different in how the same list of vocabulary was processed by students. Given this variety, I love the freedom that such an assignment offers.

Wiline's Perspective: Writing Strategies for Scientific Argument

As I realized how little writing was emphasized in my own education, I've developed a few strategies for improving the writing skills of my students. The strategies break down into four overall categories of pre-writing, drafting, editing/revising, and peer-reviewing. I develop each one below, with examples of language from class assignments.

Pre-writing: Outlining

I remember distinctly how I would write my own college papers: I would procrastinate until the week of the assignment, do all the searching of the literature and the background reading, wait and procrastinate some more. Then, a few days before the final paper was due, I would sit down and write obsessively, starting from line 1, page 1, moving through it paragraph by paragraph. Inevitably, I would go until the wee hours of the morning, struggling to finish on time. And inevitably, I would write entire paragraphs at 1:00 AM that I would end up deleting at 3:00 AM. This was, to say the least, not very efficient, nor stress-free.

The process of outlining would have spared me a lot of frustration and deleted paragraphs, but I did not truly learn how to outline until I had started my career. Outlining, if done thoughtfully, is truly the key to strong, efficient scientific writing.

When outlining, I give precise instructions to my students, and the full assignment is available on our companion website for this chapter, and it is shown in Sidebar 5.3 <https://jeremyhyler40.com/science-and-literacy/chapter-5/>.

Sidebar 5.3 Literature Review and Outline Assignment

Literature research and outline assignment

On <date>, you must turn in a typed list of <u>primary literature references</u> on your topic, as well as an <u>outline that uses these references</u>. This will be worth XX points; 25% of the grade will be deducted for each day it is late.

A. Outline

One of the things I'll be looking for in the term paper is clear organization. You may want to organize your paper chronologically, starting with the earliest historical research on your topic and moving forward to the most recent ones. You may want to organize by different techniques used to address your topic, or by different aspects of the general topic, or by different individuals or groups of people who have investigated your topic.

The important thing is that there should be some kind of clear organization, not just a random jumble of unconnected facts. We need to avoid the "laundry" list of papers summarized one after the other. I have many outlines from prior years that you can examine, housed in my office, if you'd like to see examples from other students.

When writing your outline, you can use Roman numerals for the top level, then capital letters, then numerals, then lower-case letters. You don't have to have four levels. This format is not required; you can use any outline format you want, as long as the hierarchy of topics and subtopics is clear. The lowest level of the outline should represent one paragraph in your final paper, and as such, should be <u>a fully developed topic sentence</u>. At this level, you should include some of the citations you'll use.

B. Literature Search

Your list of references should be thorough, including everything on your specific topic and all the references you used in your outline. Look at the initial three papers you found on the topic, and find every relevant reference that they cite. Also, look up some of the older papers in the Web of Science and see who has cited them. Keep doing this — for every paper you get, see who they cite and who has cited them — until you're not seeing anything new.

Your reference list will be graded based on its format (see the examples in the "Citation Format" section) and its thoroughness. If I can quickly find references that are very relevant to your topic that you haven't listed, you'll get points deducted.

<u>Your reference list must contain at least 10 references from peer-reviewed scientific journals</u>.

You may include web pages or textbooks on your list, but they do not count towards the 10 references. You may not include *Wikipedia* as a primary reference; it can be a valuable place to start looking for some kinds of information, but you should trace the information back to its original source and cite that instead. If you are having trouble finding 10, keep in mind that you'll want some more general references in addition to everything on your narrow topic. You do not have to limit yourself to articles that have your topic keywords in their title.

You will not need to use all of this initial list of references in your final paper — as you read them, you may find that some are not as relevant as the titles suggested. You may add additional references between now and writing the paper.

First, as simple as it sounds, I demand that students complete all the readings and background searching. Often, students want to create the outline, then "fill in the gaps" by reading about each header of the outline and completing a section. However, this creates a large pitfall; students write an outline before learning anything about the topic, based solely on their (sometimes uninformed) perception. Instead, by doing all the research and readings before creating the outline, they can process the information (often unconsciously) before sitting down to create the outline, thus generating an outline that is based on actual knowledge.

Figure 5.6 Outline Created by Beginning Graduate Student

a. Since their introduction into the upper Great Lakes in the early 1900s, sea lamprey populations have flourished, contributing to the collapse of important fisheries throughout the lakes.
    i. When/how they first invaded *(Source: Smith and Tibbles 1980)*
    ii. Include specific details about Lake Huron invasion since I am working with two Huron tributaries *(Source: Smith and Tibbles 1980)*
    iii. HBBS: discuss history of lamprey research in Ocqueoc and Cheboygan Rivers *(Source: Christie and Goddard 2003)*

b. In response to the sea lamprey invasion, the Great Lakes Fisheries Commission implemented several control programs responsible for reducing sea lamprey populations by about 90%, thereby allowing the reestablishment of sustainable fish stocks. *(Source: Christie and Goddard 2003)*
    i. Barriers used for lamprey control *(Sources: Lavis et al. 2003, McLaughlin et al. 2003)*
    ii. Development and use of TFM *(Source: McDonald and Kolar 2007)*
    iii. Current Sea Lamprey control methods and their downfalls (expense, blocks other valued species, etc.) *(Source: Dodd et al. 2003)*

c. An important issue facing current sea lamprey control programs is the difficulty in estimating abundance of migrating adult sea lamprey in medium and large tributaries where currently utilized mark-recapture methods for estimating abundance are ineffective. *(Source: Mullett et. al 2003)*
    i. Migration timing: trap data (downfall is that barriers are typically built several km upstream so timing estimates are inaccurate)
    ii. Abundance estimates: mark recapture methods
    iii. Emphasize that there is a knowledge gap and how the use of DIDSON can fill in this gap (This is a nice transition into the use of DIDSON for this project)

d. Frequency Identification Sonar (DIDSON) is an underwater acoustic video camera that is becoming increasingly popular in fisheries science for monitoring fish movements in rivers. *(Source: Moursund et al. 2003)*
    i. What is DIDSON?
    ii. History of DIDSON and how it has made its way into fisheries science *(Sources: Moursund et al. 2003, Baumgartner et al. 2006)*
    iii. What DIDSON has been successfully used for within fisheries science (fish counting, sizing, behavior, etc.)? *(Sources: Holmes et al. 2006, Burwen et al. 2010)*
    iv. Automated analysis of DIDSON data *(Source: Boswell et al. 2008)*

e. This study will compare DIDSON counts to the current sea lamprey index as well as provide new insight into sea lamprey migration timing, especially in comparison to other valued migrating fish species.
    i. Developing an automated program for analyzing large quantities of acoustic data *(Source: Boswell et al. 2008)*
    ii. Better understanding of their migration timing (especially in comparison to other valued species (dams; include paragraph on dams? Maybe talk about this in the paragraph regarding current sea lamprey control programs) *(Source: Dodd et al. 2003)*
    iii. Better understanding of their abundances. If not getting an exact number of migrating lamprey in the great lakes, then developing some sort of index *(Source: Mullett et al. 2003)*

f. Rationale Topic Sentence: This study will investigate whether DIDSON can provide further insight into sea lamprey migration timing as well as abundance estimates, providing a tool for evaluating the accuracy of the current sea lamprey abundance index.
    i. How accurate is the current sea lamprey abundance index? Can DIDSON give us better insight into sea lamprey abundance and migration timing?
    ii. Hypothesis I
        a. Null: DIDSON will detect no difference in migration timing between adult sea lamprey and valued migratory fish species (i.e. steelhead, walleye, sturgeon, etc.).
        b. Alternate: DIDSON will detect a difference in migration timing between adult sea lamprey and valued migratory fish species (i.e. steelhead, walleye, sturgeon, etc.).
        c. Prediction: I believe DIDSON will detect a slight difference in migration timing between sea lamprey and other migrating fish species. I think the time period in which these species are

    migrating will be very similar, however I also think their peak migrations will occur at different times and will be predicted by different environmental cues.
 iii. Hypothesis II
   a. Null: There will be no difference between the sea lamprey abundance index estimated by DIDSON and the current abundance index.
   b. Alternate: There will be a difference between the sea lamprey abundance index estimated by DIDSON and the current abundance index.
   c. Prediction: I predict that there will be no difference in the abundance index estimated by DIDSON compared to the current abundance index, which is estimated using mark-recapture methods.

References:

Baumgartner, L. J., Reynoldson, N., Cameron, L., & Stanger, J. (2006). Assessment of a Dual-frequency Identification Sonar (DIDSON) for application in fish migration studies. *NSW Department of Primary Industries–Fisheries Final Report Series.*

Boswell, K. M., Wilson, M. P., & Cowan, J. H. Jr. (2008). A semiautomated approach to estimating fish size, abundance, and behavior from Dual-frequency Identification Sonar (DIDSON) data. *North American Journal of Fisheries Management, 28.3* 799–807.

Burwen, D. L., Fleischman, S. J., & Miller, J. D. (2010). Accuracy and precision of salmon length estimates taken from DIDSON sonar images. *Transactions of the American Fisheries Society, 139.5* 1306–1314.

Christie, G. C., & Goddard, C. I. (2003). Sea Lamprey International Symposium (SLIS II): Advances in the integrated management of sea lampreys in the Great Lakes. *Journal of Great Lakes Research, 29* (Supp. 1) 1–14.

Dodd, H. P., Hayes, D. B., Baylis, J. R., Carl, L. M., Goldstein, J. D., McLaughlin R. L., Noakes, D. L., Porto, L. M., Jones, M. L. (2003). Low-head sea lamprey barrier effects on stream habitat and fish communities in the Great Lakes basin. *Journal of Great Lakes Research, 29* (Supp. 1) 386–402.

Holmes, J. A., Cronkite, G. M. W., Enzenhofer, H. J., & Mulligan, T. J. (2006) Accuracy and precision of fish-count data from a "Dual-frequency Identification Sonar" (DIDSON) imaging system. *ICES Journal of Marine Science: Journal du Conseil, 63.3* 543–555.

Lavis, D.S., Hallett, A., Koon, E.M., & McAuley, T.C. (2003). History of and advances in barriers as an alternative method to suppress sea lampreys in the Great Lakes. *Journal of Great Lakes Research, 29*(Supp. 1) 362–372.

McDonald, D. G., & Kolar, C. S. (2007). Research to guide the use of lampricides for controlling sea lamprey. *Journal of Great Lakes Research, 33* (Special Issue 2) 20–34.

McLaughlin, R. L., Hallett, A., Pratt, T. C., O'Connor, L. M., & McDonald, D. G. (2007). Research to guide use of barriers, traps, and fishways to control sea lamprey. *Journal of Great Lakes Research, 33* (Special Issue 2) 7–19.

Moursund, R. A., Carlson, T. J., & Peters, R. D. (2003). A fisheries application of a Dual-frequency Identification Sonar acoustic camera. *ICES Journal of Marine Science: Journal du Conseil, 60.3* 678–683.

Mullett, K. M., Heinrich, J. W., Adams, J. V., Young, R. J., Henson, M. P., McDonald, R. B., et al. (2003). Estimating lake-wide abundance of spawning-phase sea lampreys (*Petromyzon marinus*) in the Great Lakes: Extrapolating from sampled streams using regression models. *Journal of Great Lakes Research, 29* 240–252.

Smith, B. R., & Tibbles, J. J. (1980). Sea lamprey *Petromyzon marinus* in Lakes Huron, Michigan, and Superior: History of invasion and control 1936–78. *Canadian Journal of Fisheries and Aquatic Sciences, 37* 1780–1801.

Once the readings are done, students can sit down and organize the information they learned into a proper outline. Outlines should organize information into sections (at least two), and contain not just headers for sections, but actually go to the paragraph level. By this, I mean that each paragraph should be represented by a full sentence (the future topic sentence) that states the take-home message of the paragraph. Below each topic sentence, reading references can be indicated for further consulting (see Figure 5.6, Erin's outline for an upper level research paper, for an example; notice all the references below the topic sentences). The beauty of such an outline is that if there are logical problems in the argument, they are fairly obvious on a quick read, and paragraphs can be quickly reordered without losing a lot of work.

There is no single, correct outline, just many different ways to make an argument, as shown in Figure 5.6. But figuring out what argument the paper is telling is critical to a successful essay. Will the paper follow a historical timeline, presenting the oldest research to the newest research? Is it presenting a hypothesis, with lines of evidence for and against? Is it presenting a pattern from nature, showing evidence of the pattern in different ecosystems or organisms? Each one of these would generate a different structure.

Writing: Drafting

I make the point clear to students that once their outline is created in this intentional process, most of the work is done. Writing the actual paper just consists of filling in the gaps under each topic sentence. At this point, I suggest that each topic sentence be placed at the top of a page, so that each paragraph now has its space. I often recommend a time limit (a paragraph an hour) to break down the seemingly giant task into manageable chunks. I also propose writing in the morning, when our brains work best, and to disconnect from all distractions (i.e., turn off phones). Writing should be truly writing, not checking references endlessly.

Once we start writing a specific paragraph, I often joke that we shouldn't be too picky at first about how we write: we can always go back and fix it later. Instead, putting it all on the page without worrying about having the best sentences tend to work best. The goal is to get all the content down, not to write beautiful sentences (quite yet).

Writing: Editing

The editing part of the work is, in my experience, a critical component of writing, more so than most students expect. Once the content is down, with a sound overall structure, it's time to buckle down and look for clear, concise writing. This means, yes, sentence-by-sentence reading, slowly, even, ideally, out loud. Besides the obvious grammar mistakes to avoid, I tell my students to pay specific attention to vague terms: in a science paper, we look for precise vocabulary use. Nothing is vaguer than the noun "effect," or its matching verb "affect"; in science, everything affects everything else. In a science paper, I want to know how. In addition, we are moving away, in most science fields, from the heavy passive voice. Instead, we do take stands on positions, and I expect my students to develop clear points of view and express them. Ownership is important.

The editing work can be a daunting task for a student with little science writing experience. They may need some actual examples of good science writing, and I often show them how to transform a lengthy paragraph into a concise paragraph without losing content. I would, most of the time, work on the very first paragraph of their paper, so that they can see an application on their own work, with the intent that they can carry this over to the rest of their work. Struggling students can benefit, at this point, from the strategy of peer-review, which I develop next.

Peer Review

I have found that my skills as a science writer improved greatly once I started to read and edit another scientists' work. It seems so much easier to have a critical eye on a paper that is not our own! I now systematically make my students go through at least a couple rounds of peer-review before they turn in their final paper for a grade. I find that students gain different skills by doing such peer-reviewing: one, they receive friendly comments on their writing in a non-formal way (no grading); second, they get to see how other students have written their papers; and lastly, they can improve their own writing by editing the work of others. Often, this is a "lightbulb moment" for students in my class; they realize what works (adding headers, for instance), what doesn't work (long, confusing sentences full or jargon), and can apply these insights to their own work.

Before doing any peer-reviews, I do give precise instructions to my students, starting with the golden rule: review other people's work the way you'd like your work reviewed. Comments written in all caps, in bold, or underlined three times do not feel really good. Neither does the comment "it was great." We all have room to improve our writing, and we can't learn much from a "it was great" comment. As such, I instruct students to make several types of comments, as outlined in Sidebar 5.4.

No matter the strengths of the paper they are reviewing, I ask my students to always comment on both what needs to be fixed, and also on all the good elements to keep.

Sidebar 5.4 Peer Review Instructions

For our class, you will be anonymously reviewing the term paper of a student in the class. The purpose of peer-reviewing is to give constructive criticism that will help the author produce the best paper possible. Peer-review follows one golden rule: review and edit others' work as you would have them review and edit yours. Your peer-review will contain:

- a big-picture review;
- a small-picture review; and
- filling out the grading rubric and giving a grade to the term paper.

1. Big-picture review:

On your first reading of the paper, focus on the "big picture" of the term paper (i.e. not the writing style, etc.), specifically the following points:

- Is the topic of the paper clear, and relates to our class?
- Does the paper have major flaws?
- Is the paper well organized, and can you follow the logic of the author's arguments?
    - Do you understand the paper's overall topic and take-home messages?
    - Are important pieces of information missing?
    - Are the interpretations made by the author reasonable?
    - What is the strongest part of this paper? The weakest part?

As you do your big-picture review, write comments on the paper. You can write numbers in the margin to tell the author that you have something to say about that aspect of the paper, and then write a few sentences about that point as part of your review.

Your big-picture review can be written as short paragraphs on a separate sheet and/or as notes written in the margin of the paper.

Your review should not be condescending or insulting. Even major deficiencies in the research can be communicated in objective, unbiased, neutral terms. Be as precise in your comments as you can: a few vague comments at the end are pretty much useless for the author.

2. Small-picture review:

On your second reading of the paper, focus on more minor problems of the term paper, such as:

- Identifying minor deficiencies in the term paper, and asking the author to address them.
- Making suggestions to improve the organization of the paper.
- Making suggestions to improve the wording of sentences.
- Noting errors in punctuations and typos.

When doing this small-picture review, use the proof-reading marks indicated on the next page; it will make the communication between you and the author consistent and understandable, and it will also speed things up for you.

3. Filling out the grading rubric and giving a grade:

For this last section, fill out the grading rubric provided on the last page of this handout to the best of your capacity. Some areas will be easier to assess for you than others, but that's alright, do the best you can. Justify where the author lost points as much as you can (and in the same way you would like to be graded by instructors). The more precise you can be on where the author lost points, and how many points were lost for given problems, the easier it is for the author to see where they need to spend some time revising. Then add up the scores from the rubric, and give the term paper a grade.

4. Remember the golden rule of peer-review:

Review and edit others' work as you would have them review and edit yours!

I love running multiple rounds of reviews, switching out the pairings of students. I typically pair students once based on topic, so that students get one reviewer that may know something about the topic of their paper. Then I pair students based on their writing strength, so that they may further improve their writing. I also like to pair strong writers with struggling writers for at least one round of peer-review. I find that both types of writers gain a lot from these interactions: strong writers get to help peers, while struggling writers get to see what the benchmarks are for such assignments.

Literature that Uses Science

For both Wiline and Jeremy, making connection to literature is one additional piece of their inquiry-based teaching. However, it can be difficult for teachers to find fiction texts that actually incorporates science and scientific concepts. More importantly, it is difficult to find fiction texts that accurately depict science. Below is a list of books with brief explanations that have the potential to be useful assets in a science classroom.

- *Hatchet* by Gary Paulsen (2009) — This text could be used by taking a different view of the main character Brian in the story. If students take the perspective of Brian being an invasive species, then students can study the impact Brian has on the animals and environment where he is lost. Then, this can lead into scientific concepts of invasive species native to the state they live in and even our own country. Additionally, *Hatchet* introduces students to the diversity of life present in this type of ecosystem, including not just animals, but also plants.
- *My Side of the Mountain* by Jean Graighead George (2004) — This book can be used in the same type of fashion as *Hatchet*: the text describes, in great detail, the type of animals and plants found by the main

character, including the ones that allow him to survive. Reading this book brings Wiline right back to her ethnobotany classes, where she was asked to actually taste wild fruits!
- *Hoot* by Carl Hiaasen (2005) — A great book to help students understand human impact on ecosystems and animals. Besides having a great narrative story line, there are many great scientific talking points to infuse into the science classroom dealing with how we as humans don't consider the consequences to our economic actions. For instance, as the construction company continues to push forward — even when they knew that the owls were living on the site — illuminates many issues related to human impact, endangered species, and encroaching human development.
- *Pax* by Sara Pennypacker (2019) — A great book that takes the first-person narrative approach for both a boy and his pet fox, Pax. While the story line is engaging, the natural history of foxes is also accurate — the combination of the two makes it a wonderful text that could be easily tackled in both ELA and Science simultaneously.

Also, please note that the National Science Teachers Association releases, annually, lists of Outstanding Science Trade Books for Students K-12, and a link to those lists is available on our companion site.

Apps, Websites, and Devices Worth Considering in the Science Classroom

There is no way that we can create an exhaustive list of apps, websites, and devices that are worth your time to explore and — even if we did — it would be out-of-date by the time we publish this book.

To the best of our ability, we will keep an active list of such resources up to date on this chapter's companion page. That said, there are a few resources that we feel are definitely worth your attention, and we provide a brief, annotated list of them here. Additionally, with any of the technologies that we mention here, we note that it is important to take an inquiry stance, to invite students to be creators (not just consumers), and to explore connections to the ISTE standards.

In particular, as these resources relate to the ISTE Standards for Students (2016), we encourage educators to focus on the "Knowledge Constructor," "Innovative Designer," "Computational Thinker," "Creative Communicator," and "Global Collaborator" standards. We list the main standard for each of these here as a way to frame this final section:

- Knowledge Constructor: "Students critically curate a variety of resources using digital tools to construct knowledge, produce creative artifacts and make meaningful learning experiences for themselves and others."
- Innovative Designer: "Students use a variety of technologies within a design process to identify and solve problems by creating new, useful or imaginative solutions."
- Computational Thinker: "Students develop and employ strategies for understanding and solving problems in ways that leverage the power of technological methods to develop and test solutions."
- Creative Communicator: "Students communicate clearly and express themselves creatively for a variety of purposes using the platforms, tools, styles, formats and digital media appropriate to their goals."
- Global Collaborator: "Students use digital tools to broaden their perspectives and enrich their learning by collaborating with others and working effectively in teams locally and globally."

With everything shared here, we encourage readers to view all of the ISTE standards as a way to design effective, engaging lessons with the resources shared below. This is the kind of thinking that high-quality science and literacy instruction should foster, and the technologies shared here can lead students towards those goals.

First, we share resources that we have found helpful when exploring ideas in the NGSS, pushing us to think of scientific phenomena that can be shared at the beginning of a new unit of study. As a reminder, the NGSS defines phenomena as "observable events that occur in the universe and that we can use our science knowledge to explain or predict" (Achieve Inc., Next Generation Science Standards, Next Gen Storylines, & STEM Teaching Tools, 2016). The document then articulates the importance of this approach (emphasis in original):

> By centering science education on phenomena that students are motivated to explain, the focus of learning *shifts from* **learning about** *a topic to* **figuring out** *why or how something happens*. For example, instead of simply learning about the topics of photosynthesis and mitosis, students are engaged in building evidence-based explanatory ideas that help them figure out how a tree grows.

Introducing phenomena to students can be a challenge, both in terms of shifting our perspectives on teaching and in finding high-quality examples. To that end, three different resources have been particularly helpful (links available on companion page for this chapter):

- From the NGSS, the resource mentioned above can be found on their main website for discussing the use of phenomena in science, and is available under a Creative Commons license. Additionally, that page links to other resources from the Research and Practice Collaboratory as well as a video from the Teaching Channel.
- Maintained by TJ McKenna, a professor at Boston University focused on Science Education, the *Phenomena for NGSS* website includes a searchable database of dozens of examples, each with images or video clips. Similarly, the *Wonder of Science* website created by 2011 Montana Teacher of the Year Paul Andersen has dozens of examples, sorted by grade level and available as a Google Doc with links.
- Finally, Karla from *Sunrise Science* (2018) has collected a number of additional links that lead to even more websites and organizations focusing on phenomena in her post, "Free Websites for Finding NGSS Anchoring Phenomena."

Second, we share resources that provide unique insights on scientific concepts, especially biological and ecological concepts, that require much more than a quick glance. These websites and apps can be used as the foundation for an entire unit of study, with multiple opportunities for students to explore the rich resources within:

- Identifying that Plant, created by plant aficionado Angelyn Whitmeyer, offers a number of its own resources for plant ID, including a gallery of plant portraits, and also links out to numerous other plant ID websites.
- OneZoom, a charitable organization from the UK, has created the interactive Tree of Life Explorer, "[a]n interactive map of the evolutionary relationships between 2,123,179 species of life on our planet," inviting users to see unique connections in a way that only interactive media can provide.
- The Concord Consortium offers teachers and students "scientifically accurate virtual labs and hands-on digital tools." Available for free, the vast collection of resources is easily searchable, and teachers can create a class dashboard by signing up for a free account.

Third, as we continue to think about new ways to bring creative nonfiction and writing-to-learn strategies into our classrooms, we share a few additional places to find strategies that can be adapted quickly for multiple contexts.

- *The Writing Fix*, a site created by a writing teacher and former National Writing Project site director, Corbett Harrison, contains countless

lessons, including an entire section on general strategies for writing across the curriculum, with additional ideas for science, math, and other content areas.
- As a long-time favorite of writing teachers, Barry Lane has been encouraging students to use their voices, making nonfiction writing both a creative and engaging process. His book (and companion website), *51 Wacky We-Search Reports: Face the Facts With Fun* (2003), offers a variety of fun ideas. In addition, he has made many of the strategies available for free in a PDF handout, also linked on this chapter's companion page.
- Even though it is only available as a PDF, the Michigan Department of Education produced a number of writings across the curriculum guidebooks that are filled with many ideas. The *Writing Across the Curriculum: Science* guide, co-authored with the Michigan Science Teachers Association, is available as a link, too.

Again, there is no way that we can accurately capture all of the great resources available online, so we conclude the chapter here with the reminder that this chapter's companion page on the website will be the best place to go for up-to-date links and information.

Conclusion: Foreground Literacy Practices During Inquiry Activities

In summary, all of these strategies and technologies can enhance the integration of science and writing, yet only when used with intention. Although this chapter has an assortment of suggestions — some of which could fill a single class session and others that could stretch out over days — it is sometimes difficult to know exactly what to do, when to fit it in to an inquiry-based unit, or how much time to take.

Though it may sound trite, our advice is simple: trust your gut.

Perhaps it is a Friday, and you want to save the final test or student presentations until Monday; that might be a good time to have them create a unit review via a concept map. Perhaps you just introduced a difficult concept the day before, and students are still struggling; that might be a good day to have them try to describe the concept using a unique genre like a recipe, menu, or even an order of service for a church gathering. And, perhaps it is the middle of a unit, and students really need to step back from what they are doing with a mountain of vocabulary words, and having a chance to play with those words through creative nonfiction might be just what they need for a day.

Each activity in this chapter is flexible, both in terms of how much total time you want to devote to it as well as to when, during a unit of instruction, you might implement it. The point of any of these, however, should be clear: integrating active literacy practices into inquiry-based learning helps students solidify what they know, as well as how they know it. Our hope is that you will find creative ways to integrate these approaches and tools into your daily lessons, lab experiences, and differentiated assessment options.

6

Professional Development in Science and Writing

Professional learning opportunities for teachers of science have, in the past few years, revolved more and more around the Next Generation Science Standards. This has led, as we've noted throughout the book, to many trends that include a focus on inquiry, the integration of scientific phenomenon, and more intentional uses of writing. The cross-cutting themes in the NGSS, too, provide teachers with new ways of thinking about curriculum and organizing instruction. While all change, especially change in educational practices, takes time, we are pleased to know that the NGSS have moved teachers in new directions.

So, how do we plan for professional learning in science, accounting for these changes and building on best practices? As we have continued to think through these changes and challenges, the three of us have collaborated each of the past four summers to create an opportunity for science and literacy teachers to learn together over the course of a one-week, intensive, place-based program. The Beaver Island Institute provides teachers with the time, space, and experiences to rethink their approach to writing in science by immersing them in scientific inquiry and leading them through a variety of approaches to teaching writing.

Building on the mantra of "teachers teaching teachers" that is a prominent component of professional learning in the National Writing Project, the three of us have worked with our colleague Jeremy Winsor (and, more recently, with Merideth Garcia and Courtney Kurncz) to create a structure for the week, offer specific lab experiences in the field, and help participants develop their own units of study that they can take back to the classroom. Thus, this chapter provides some background on the rationale for the structure of the institute, as well as some of the activities and outcomes we have seen emerge from it.

The Design of the Beaver Island Institute for Science and Literacy Teachers

As Maria Gigante argues in her article "Navigating the Next Generation Science Standards: Implications and Implementation for Faculty in Writing and the Sciences," there are many challenges that science teachers face, and layering additional opportunities for writing makes the work even more complex. She asks: "How can writing be positioned as integral to science education rather than as a risky importation that could disrupt science teaching and learning?" (2016, p. 95). As we have argued throughout this book — and in the design of the institute itself — we argue that helping students write as disciplinary experts should be our goal. We share our model for the institute here as a way for our readers to envision a similar kind of professional learning structure that could be adapted to local needs and contexts. When promoting our institute to teachers, we invite them with the message in Sidebar 6.1.

When Wiline and Troy began planning the first institute in 2016, we looked to her experience as a biology instructor who had offered many courses to undergraduates involving field-based laboratories in the island's

Sidebar 6.1 Language from the Advertisement for Our Beaver Island Institute

> We welcome ELA and Science teachers — ranging from upper elementary to lower high school grade levels — to experience a unique, one-week professional development institute at CMU's Biological Station on Lake Michigan's Beaver Island. Study ecosystems, educational technology, and strategies for integrating reading and writing into the classroom.
>
> Working in pairs, this workshop is designed for science and ELA educators based in the same middle school. Teachers will study ecosystems, technology, and strategies for integrating reading, writing, critical thinking into the classroom. Participate in:
>
> - Individual and small group work.
> - Field activities to promote discussion and collaboration.
> - Science and literary activities.
> - Collaborative, inquiry-based units identifying key standards from the Next Generation Science Standards, Common Core Literacy Standards, and ISTE (technology) standards.
> - Integrative teaching practices to integrate mobile technologies (smartphones and tablets, as well as apps) into teaching and as a learning tool for students.

wetlands, dunes, forests, streams, and out on the waters of Lake Michigan. We envisioned an immersive professional development institute that would integrate the study of various experimental designs (notably observations, natural and field experiments) as a means of combining reading, writing, and critical thinking into the classroom.

Troy's work with the National Writing Project (NWP) informed the daily patterns of the institute, the arc of the week as a whole, and the final deliverables asked of teachers. In their list of core principles, the NWP contends that "[e]ffective professional development programs provide frequent and ongoing opportunities for teachers to write and to examine theory, research, and practice together systematically" (National Writing Project, n.d.). Thus, by engaging in the writing process, in the discipline of science, we invite teachers to rethink their own pedagogy.

Put another way, Carter, Ferzli, & Wiebe frame the opportunity in an additional question:

> But what if faculty in the disciplines reconceived the goal of education as helping students to grow as participants in the ways of knowing of their disciplines? Students would be understood not as outsiders but as fellow members of the disciplinary community, albeit on its periphery. Teaching would be understood as creating opportunities for students to learn by doing the kinds of activities full members do, though in a form appropriate for apprentices. And writing would be understood as the critical link between doing and knowing in the disciplines.
> (Carter, Ferzli, & Wiebe, 2007, p. 299)

The workshop has been designed for educators to work in pairs, although we often accepted individual educators, too. By the end of the week, pairs are asked to create a rough draft of a unit plan that they could use during the following school year.

In particular, on Friday morning, they participate in a protocol called a "charette," where they present their work in progress and receive feedback from the group. The charette protocol, designed by educators with the School Reform Initiative, "is a term and process borrowed from the architectural community. Its purpose is to improve a piece of work" (Juarez, Thompson-Grove, & Feicke, 2007, p. 1). This process invites them to share their "work in progress" for about five minutes, stating explicitly the kinds of feedback that they are interested in receiving before they begin. Once completed with the presentation, the teachers remain quiet as they listen to questions, comments, and suggestions from their peers. Finally, after five minutes of peer response, the presenting teachers are able to then share their ideas about the next steps for the work.

Given that the primary goal of the Beaver Island Institute was to help teachers reconsider the design of science and literacy activities in light of the Next Generation Science Standards, the Common Core State Standards for Literacy, specifically related to argument writing and nonfiction reading, as well as the ISTE standards for integration of technology, we share the following goals with participants:

- Identify key standards from the Next Generation Science Standards, Common Core State Standards in literacy, and ISTE standards for educational technology to design a collaborative, inquiry-based unit with their teaching partners.
- Engage in a variety of science and literacy activities as learners, then analyze and reflect on their experience to create a series of appropriate lessons, labs, and activities for their unit.
- Integrate mobile technologies (smartphones and tablets, as well as apps) into their teaching practice and as a tool for students to document their own learning.

In order to accomplish these goals, we created a flexible structure for each day, typically organized as shown in Sidebar 6.2.

Sidebar 6.2 Beaver Island Institute Structure

Daily Schedule

- 7:15 — Breakfast
- 8:15 — Writing into the day
- 9:00 — Field activities to build scientific content knowledge
- 12:00 — Lunch
- 1:00 — Independent work time with partners
- 2:00 — Afternoon science and literacy activities
- 5:30 — Dinner
- 6:15 — End-of-day wrap up and reflections

Over the course of the week, we ask teachers to engage in the process of learning as if they were students, but periodically to pause, reflect, and consider implications for their own classroom. Writing prompts for each morning have changed a bit over the years, but are generally based on the following themes:

- **Sunday** afternoon and evening: travel from Charlevoix on the Beaver Island Ferry; arrival on the island and transport to the station; ice breaker activities relating to the nature of inquiry-based learning in science and literacy.
- **Monday**: field trip to wetlands for observational study of snake morphology and dispersal patterns; afternoon work analyzing data collection from the morning and Gyotaku (Japanese-style) print making; individual work time for project; reflection and debriefing the day.

- **Tuesday**: field trip to two local streams to collect macroinvertebrate specimens and test water quality; afternoon work analyzing data collection from the morning and participation in "Lazy Lizards" (natural selection simulation); individual work time for project; reflection and debriefing the day.
- **Wednesday**: field trip to beachfront/dune to pull invasive species (spotted knapweed), preparing site for long-term ecological study; land art exercise using pulled knapweed; afternoon work analyzing data collection from the morning and participation in art and writing activities to stimulate creative thinking; individual work time for project; reflection and debriefing the day.
- **Thursday**: field trip aboard the research vessel *Chippewa*, with dredging of Lake Michigan to collect soil samples; writing-to-learn activities including explanations of charts and graphs as well as exploring creative nonfiction; individual work time for project; reflection and debriefing the day.
- **Friday**: participants shared their work in progress using a "charette protocol," in which they talked about their projects for five minutes, received three minutes of feedback from the group, and discussed next steps for their work in two minutes; final institute evaluation and preparation for departure.

Many of the additional writing strategies have been documented throughout the book, even though we reiterate them briefly here. While we do not have space to explore each of our daily activities in detail, we describe five types of experiences that we do on the island as ones that could, with some creativity and flexibility, be adapted to many local contexts.

The Field Experiences

Given our goals for creating an inquiry-based, writing-rich experience for participants that they could adapt to their own teaching contexts, there are a number of experiences that we do throughout the week. We describe five of them, briefly, and in the next section, describe ways that we invite teachers to wrestle with the data that they have collected.

Field Experience 1: Snake Boards (Observational Design)

Purpose: Introduce participants to observational studies, field data collection as a team, and inquiry-based data handling.

Data to be collected: In the field, as individuals (splitting up boards between the whole group) for the team, participants observe the number of species, the frequency of particular species as compared to one another, and dispersal patterns.

Scientific objectives:

1. Explore the potential of observational studies in sciences.
2. Develop two observational studies in ecology, behavior and morphology.
3. Become familiar with dispersion patterns of organisms.
4. Witness antipredator behaviors of several snake species.
5. Correlate antipredator behaviors with morphological adaptations.
6. Handle data collection and data handling to answer biological questions.

Engagement factors: full-morning field trip, high energy (or, perhaps, stress!) activity searching for the snakes, as a first attempt at collecting data, it can be used as a springboard/touch point for rest of the week.

Positive aspects of this type of observational studies: low expense (no need for fancy equipment), unless field trip requires busing; high relevance (observing real ecological patterns in nature).

Potential drawbacks of this type of observational studies: low control over what we will find on a specific day, and on actual results collected (as such, activities such as this are fairly variable in their success); can be highly affected by weather conditions.

NGSS covered by activity:

- 3-LS3-2: Use evidence to support the explanation that traits can be influenced by the environment.
- HS-LS2-8: Evaluate evidence for the role of group behavior on individual and species' chances to survive and reproduce.
- 4-LS1-2: Use a model to describe that animals receive different types of information through their senses, process the information in their brain, and respond to the information in different ways.
- HS-LS2-2: Use mathematical representations to support and revise explanations based on evidence about factors affecting biodiversity and populations in ecosystems of different scales.

As a relatively reliable field observational study, on the first day of the institute we head out to large fields to check out the snake boards. Beaver Island is famous for its snake diversity (luckily, all of them non-venomous), and this activity utilizes the natural history of the island, combined with the resources provided by CMU's Biological Station. Scattered throughout a couple of CMU fields, about a hundred plywood boards provide ideal habitats for the large diversity of snakes present on Beaver Island. After a quick debrief on snake identification and retreat behaviors, participants walk out with notebooks in hand, prepared to lift the boards and search for various species of snakes (Figure 6.1). They note not only whether they find snakes, but also what species and their escape behaviors.

Figure 6.1 Participants Lifting Snake Boards During the 2019 Beaver Island Institute (Photo by Jeremy Winsor)

Because this is an observational study, the results can be hit or miss. Most times, we are able to find a number of snakes under at least half of the boards, establish the type of dispersion pattern followed by the snakes (clumped dispersion), and correlate some escape behaviors to morphology.

We will describe this initial data collection activity from the Snake Board experiment in more detail in the next section of the chapter, "Capturing and Interpreting Data: A Closer Look."

Field Experience 2: Line Transects/Botany

Purpose: Introduce participants to natural experiments (pros, cons), field data collection as individuals, introduce idea of notebooking/drawing, graphing data.

Data to be collected: Identify plants and collect abundance data.

Scientific objectives

1. Explore the potential of natural experiment studies in sciences.
2. Carry out a natural experiment study in ecology.
3. Become familiar with plant diversity along the Michigan dunes.

4. Witness an example of ecological succession.
5. Handle data collection and data handling to answer biological questions.

Engagement factors: Field trip; plants can be fun to examine, time on the beach.

Positive aspects of this type of natural experiment studies: low expense (no need for fancy equipment), unless field trip requires busing; high relevance (observing real ecological patterns in nature).

Potential drawbacks of this type of natural experiment studies: activity relies on finding differences in nature large enough for participants to pick out; can be highly affected by weather conditions.

NGSS addressed by activity:

- 2-LS4-1: Make observations of plants and animals to compare the diversity of life in different habitats.
- HS-LS2-2: Use mathematical representations to support and revise explanations based on evidence about factors affecting biodiversity and populations in ecosystems of different scales.

For a second type of data collection that uses variations found in nature, we visit the beach. In particular, we visit a beach that — over the span of about 200 yards — demonstrates ecological succession quite well, from the bare shoreline to the tree line. About every 10 meters, participants are asked to set up a one-square meter plot and identify the plants that are present in the plot. Participants then continue to sample the dune's flora at multiple spots along the transect to capture the change in species composition (Figure 6.2).

By counting the total number of species in each plot (species diversity), participants can see patterns of biodiversity emerge from the shore to the tree line. While they often predict that biodiversity of plants in any plot would increase as they near the tree line, the data shows that diversity is highest in the middle of the dunes (halfway between the shoreline and the tree line), which is a well-demonstrated scientific phenomenon, the "intermediate-disturbance hypothesis."

This process is both systematic and enlightening. Having had the experience of collecting what essentially turns out to be random data about the snakes the day before, participants are much more attuned to the types of data they will need to collect. Also, knowing that there are other teams going through the same process and collecting the same data, they understand that their data will be compared and aggregated.

While dunes might not be easily accessible for most school districts, similar experiences can be done with any old agricultural field left alone for more than a few years. Often, the transitions from field to mature forest are visible on the edges of such plot of lands.

Figure 6.2 Participants Collecting Flora Data During a Dune Transect During the 2017 Institute (Photo by Troy Hicks)

Field Experience 3: Stream Sampling

Purpose: Introduce participants to natural experiments, field data collection as individuals, combining data as a group, introduce idea of notebooking/ drawing, graphing data.

Data to be collected: Collect stream macroinvertebrates, identify species of macroinvertebrates, and infer health of stream from data.

Scientific objectives

1. Explore the potential of natural experiment studies in sciences.
2. Carry out a natural experiment study in ecology.
3. Become familiar with macroinvertebrate diversity in streams.
4. Identify macroinvertebrates using dissecting microscopes and dichotomous keys.
5. Handle data collection and data handling to answer biological questions.
6. Graph class data.

Engagement factors: Field trip; handling live animals; getting wet!

Positive aspects of this type of natural experiment studies: Low expense (no need for fancy equipment), unless field trip requires busing; high relevance (observing real ecological patterns in nature).

Potential drawbacks of this type of natural experiment studies: [same as Field Experience 2] activity relies on finding differences in nature large enough for participants to pick out; can be highly affected by weather conditions.

NGSS addressed by the activity:

- 2-LS4-1: Make observations of plants and animals to compare the diversity of life in different habitats.
- HS-LS2-2: Use mathematical representations to support and revise explanations based on evidence about factors affecting biodiversity and populations in ecosystems of different scales.
- HS-LS2-7: Design, evaluate, and refine a solution for reducing the impacts of human activities on the environment and biodiversity.

Every year, we've run another type of natural field experiment that matches the same objectives and purposes as the Field Experience 2, and continues to be quite fun to do. We ask participants to sample two different streams that vary greatly in their quality; one is in a pristine area of the island, a little trout creek that runs through a forest ecosystem, and the other is in a heavily disturbed area with lots of human activities. Participants get to use different types of nets and buckets to collect macroinvertebrates at both locations (Figure 6.3). Following the sampling effort, we bring back all of the specimens to the classroom to identify the major groups of invertebrates present using a dichotomous key and dissecting microscopes. Based on what the participants find in both locations, we then compare biodiversity between the two streams.

This activity generates a lot of excitement and can be easily replicated at home with students in any local streams. If local streams are hard to reach for students, macroinvertebrates can be collected fairly easily by instructors and brought to the classrooms.

Field Experience 4: Trail Cameras

Purpose: Introduce participants to trail camera technology, picture-base data sets, and whiteboarding.

Data to be collected: Tallies from screenshots.

Scientific objectives:

- Explore the potential of observational data.
- Explore a picture-base dataset.
- Answer a biological question of choice.
- Represent data graphically.

Figure 6.3 Participants Sampling for Macroinvertebrates During the 2018 Institute (Photo by Troy Hicks)

Engagement factors: Freedom to explore different questions; can look at megafauna; always fun to see what happens nocturnally.

Positive aspects of this type of observational studies: High relevance (observing real ecological patterns in nature); high excitement for students to look through pictures, not knowing what they will find; independence in deciding what to focus on.

Potential drawbacks of this type of observational studies: Initial cost of buying the trail cameras; you never know what students will find; students can have a hard time deciding what to concentrate on given this much freedom.

NGSS addressed by the activity:

- 2-LS4-1: Make observations of plants and animals to compare the diversity of life in different habitats.
- HS-LS2-2: Use mathematical representations to support and revise explanations based on evidence about factors affecting biodiversity and populations in ecosystems of different scales.

Both to have some extra data in our back pockets in case of a rainy day, as well as to show another aspect of technology use in data collection, our colleague Jeremy Winsor has each year set up trail cameras to capture activity near the station's compost pile. A healthy walk into the woods is required to get to the compost pile, and there are many points where trail cams can be set up to capture wildlife including deer, raccoons, opossums, sea gulls, and sometimes even a coyote or a bald eagle.

Cameras are set up for at least 24 hours, sometimes more if we can. Winsor then downloads select snapshots of all the videos collected, creating a folder in Google Drive with about two dozen pictures across the 24 hours. Admittedly, he is selective, not taking a random sample, but instead ensuring that the images have some identifiable wildlife in the frame. This then allows everyone to view the images, noting the time and the number of particular species present.

While not quite as systematic as it could be, we have to balance the amount of time we want participants sorting through photos and video clips with the overall goal of creating charts that represent species diversity over the course of 24 hours. If they were to sort through every minute of every video, that could take hours of workshop/class time. Thus, as the teacher, setting up an activity like this will likely take some judicious curating of existing material.

Still, even this way, representing the relative frequency of particular species in contrast to others (like seagulls and raccoons as compared to coyotes) is very useful. As described in Chapter 4, this activity is well-suited for whiteboarding (see Figure 6.4) and participants are asked to represent their data in at least two different ways.

Field Experience 5: Removal of Plant Invasive Species

Purpose: Get participants involved in the removal of plant invasive species and evaluate the biodiversity of plants before the removal and after (long-term study over several years).

Data to be collected: Diversity of flowering plants within an area, amount of a specific invasive species removed (number of plants and weight).

Figure 6.4 Participants Whiteboarding the Trail Camera Data During the 2018 Institute (Photo by Troy Hicks)

Scientific objectives:

1. Explore the potential of experimental studies in sciences.
2. Develop a research question, hypothesis and prediction.
3. Carry out an experiment, with proper replicates and controls.
4. Become familiar with chipmunk behavior.
5. Handle data collection and data handling to answer biological questions.
6. Process information in a visual manner.
7. Write a conclusion regarding a dataset.

Engagement factors: Working outside, and pulling invasive species (which is always memorable and highly satisfying).

Positive factors of this type of field work: Satisfaction of pulling an invasive species, with the feeling of helping conservation efforts, even if only a little bit; potential for long-term studies to see if native flowering plants increase in abundance after the removal of invasive species.

Drawbacks of this type of field work: Highly dependent on season, especially as it's much easier to identify flowering plants in late spring or summer.

NGSS Addressed by This Activity:

- 2-LS4-1: Make observations of plants and animals to compare the diversity of life in different habitats.
- HS-LS2-2: Use mathematical representations to support and revise explanations based on evidence about factors affecting biodiversity and populations in ecosystems of different scales.
- HS-LS2-7: Design, evaluate, and refine a solution for reducing the impacts of human activities on the environment and biodiversity.

Another powerful field activity is to involve participants in the removal of invasive plant species. At the Biological Station, spotted knapweed (*Centaurea maculosa*) is an invasive plant that is quickly taking over the dune ecosystem. As part of this activity, participants first establish a list of all of the species currently in a large quadrat (in our case, a 25x50m rectangle). This can take some time, as some species may be difficult to identify and require some internet searching, field guides, or reliance on expert botanists. Following the identification of all plants, participants then spread out on the plot and pull all spotted knapweed, keeping track of the number of plants they are pulling (Figure 6.5). At the end, we weighed the knapweed to get an idea of the total biomass we've removed. The goal of such removal is to then watch, over the next five to 10 years, the native species bouncing back, with new species appearing in the plots. Though they will not physically see the results of their efforts unless they some day return to the island, participants take great pride in beginning work on a long-term study.

Participants can take this type of activity to any areas around their school, especially as invasive plants thrive in disturbed habitats such as the edges of roads or parking lots. There is something both inherently satisfying and emotional in pulling large quantities of an invasive species: for once, we feel like we are helping out, even though just a bit. It can be a great way to address environmental issues with our middle schoolers, with a positive activity linked to it.

Figure 6.5 Participants Pulling out Spotted Knapweed During the 2019 Institute (Photo by Troy Hicks)

Field Experience 6: Animal Behavior

Purpose: Introduce participants to field experiments (pros and cons) and experimental rationale, working with wild animals, and continue to build on data handling skills.

Data to be collected: Behavioral data (frequency and duration of specific behaviors).

Scientific objectives:

1. Explore the potential of experimental studies in sciences.
2. Develop a research question, hypothesis and prediction.
3. Carry out an experiment, with proper replicates and controls.
4. Become familiar with chipmunk behavior.
5. Handle data collection and data handling to answer biological questions.
6. Process information in a visual manner.
7. Write a conclusion regarding a dataset.

Engagement factors: Working with live animals such as birds or chipmunks, observing behaviors outside.

Positive factors of this type of experimental studies: Hands-on experience in watching wildlife and applying the scientific method.

Drawbacks of this type of experimental studies: Sometimes animals don't cooperate; weather can determine the success of this type of study.

NGSS addressed by the activity:

- 2-LS4-1: Make observations of plants and animals to compare the diversity of life in different habitats.
- MS-LS1-4: Use argument based on empirical evidence and scientific reasoning to support an explanation for how characteristic animal behaviors and specialized plant structures affect the probability of successful reproduction of animals and plants respectively.

Finally, as one of the activities that Wiline runs with her own students over the course of many days, participants get a taste of a true field experiment by watching the foraging patterns of Eastern chipmunks. Around the station, many burrows are already marked with small flags, and Wiline walks groups around to find particular burrows to observe. Chipmunks are asocial and, as such, do not share a burrow. By watching a specific burrow, we can then be fairly certain of the identity of the chipmunk without the hassle of trapping it.

Before heading out to watch chipmunks, Wiline shares a number of different types of variables that we could introduce into the experiment.

Because chipmunks stockpile food resources into their burrows, they readily participate in food-based experiments that involve sunflower seeds: if we put a pile of sunflower seeds nearby, chipmunks will take trips back and forth between their burrows and the food pile, filling up their cheek pouches, to carry all the food possible back home.

This creates an ideal experimental set-up: if we know that a chipmunk will always come back to a spot to carry back seeds, we can manipulate the feed pile in a number of different ways to see if we see a behavioral response from the chipmunk. That is, if we make the food pile "scarier" (out in the open, far from cover or refuge), will the chipmunk stay as long and come back as often? What if we play a sound of a predator? What if we make the food pile "lousier" (now we mix sand and dirt with the sunflower seeds so that they are harder to get)? What if the food pile is further away from home? Next to a very aggressive chipmunk? What happens if we put the food pile inside a maze? How quickly can they solve the maze? By playing with these variables, participants begin to understand the aspects of experimental design.

Depending on the year, participants have had the opportunity to choose a specific research question, design an experiment to answer the question, and explore topics such as controls, replicates, and confounding variables.

We've had varying levels of success with this particular activity, depending mostly on weather (chipmunks don't like to participate in any experiments if it's rainy or windy). But the experience of watching animals in the wild, in an experiment, is both memorable and rewarding (Figure 6.6). Also, if we were to do this experiment with different kinds of animals — say birds,

Figure 6.6 Participants Watching Chipmunks Foraging at an Experimental Feed Pile During the 2018 Institute (Photo by Troy Hicks)

snakes, or field mice — it would certainly take a lot more patience and attention. Chipmunks are ideal candidates because we know, at least approximately, the time and places that they will feed. An activity like this could be adapted to a local context, and there are examples of animal spotting citizen science projects such as Project Feeder Watch, linked on the companion page for this chapter.

Capturing and Interpreting Data: A Closer Look

With each activity, we also invited teachers to gather data. Depending on the activity, the data collection varied, and teachers would initially collect their data in written notebooks while out in the field. Once returning to the biological station, we would then open up a shared Google Spreadsheet and ask teachers to begin inputting their results. As a seemingly simple task, the first time we do this activity in each institute, it has been one of the most engaging — and simultaneously most frustrating — events that the teachers experience.

As noted above, the Snake Board activity (which is approved by CMU's research ethics committee) is the first activity where teachers engage in data collection. Before heading out into the field, teachers have been introduced to the snakes' morphology, the fact that different species have different dispersal patterns, and that they may find some boards with multiple snakes, some with only one, and some with none at all. Rather than sharing a method for collecting and organizing the data before we get into the field, we encourage participants to talk to each other and decide how to collect data (but we do not demand that they have a plan in place). Thus, when we reach the field, teachers are working with their partners and describing what they see, sometimes with tally marks, in writing, or even with pictures. Some even draw a map of the wetland, and the approximate locations of the snake boards. Some do none of the above, and get completely absorbed in the task of finding as many snakes as possible!

Upon returning to the station, we then ask the teachers to begin organizing the data in a shared Google Sheet. They start creating columns and rows, labeling the different kinds of snake species that were found, as well as frequency. Soon, they start questioning why and how the data are organized and, ultimately, as Wiline would state it, "What story the data is telling." This process has been, with each group, both an infuriating and enlightening exercise. As you can see from the looks on some of the participants faces in Figure 6.7, the group was engaged in intense conversation about how to organize the data. While they worked to collaboratively enter the data into a Google Spreadsheet, they were drawing information from their field notes as well as an initial brainstorming activity that they had done on the whiteboards. Needless to say, the conversation was intense, yet ultimately generative, and the teachers understood that the productive struggle was, indeed, worthwhile as they moved through their inquiry.

This process forced us, as facilitators, to consider when, why, and how much instruction we provide for our participants before engaging in the process of data gathering, organizing, and analysis. On the one hand, we concur with the participants. A little more direction, especially with this first field experience and before leaving the station in the morning, could have been helpful. They would have some context for what they are looking for, as well as how they could work collaboratively to be more efficient in the process of doing so. The process of data collection in the field as well as organizing the data on the spreadsheet would be much smoother and could be plugged into a pre-formatted template for creating a chart.

On the other hand, taking a move from Wiline and the ways that she structures the activity for her undergraduates, we chose not to do so. By creating a space for cognitive dissonance, by consciously staying away from providing a protocol for participants to collect data in a structured manner, she sets them up for a more significant lesson: the need to create a research question, tailor data collection to that question, and be deliberate in the process of

Figure 6.7 Participants Figuring Out How to Process the Data During the 2018 Institute (Photo by Troy Hicks)

gathering that specific type of data. Despite their initial frustration, Wiline's undergraduates — as well as the participants in our institute — come to see the method to our madness. Although their initial frustration may have been a bit overwhelming, the point of the activity becomes clear.

Later in the week, as the experiments continue, participants do make more intentional plans about data collection and, ultimately, analysis. Connecting to our broader goals of integrating different kinds of modeling activities, genres of writing, and opportunities for technology integration, this moment of initial frustration with the snake board data on Monday morning serves as a touchstone throughout the week, helping teachers then think about the ways they can invite students to pose their own critical and creative questions, thus leading to a more organized data collection process.

In short, we want participants to have some productive struggles. Much like their students would be expected to do in their own classrooms, the teachers in the institute need to experience the feeling of frustration as a way to — in future experiments — ensure that they are making smart decisions about the questions they ask, the data they collect, and how to best organize and represent these data in their findings.

It is through these types of field-based experiences that the teachers come to understand the inquiry process, thinking critically and creatively about the types of data they could ask their own students to collect, over time and in particular settings. More importantly, it gives them opportunities to manage real data, not just abstract data sets or examples from textbooks. Much like there is value in cooking from scratch or making crafts by hand, there is also value in collecting one's own data and translating it into a scientific argument.

Outcomes and Implications

From our four years of experience with Beaver Island Institute participants, we have found a variety of outcomes based on anecdotal interactions with participants, an end-of-institute survey, and formal interviews and classroom visits.

Participant Responses via Institute Surveys

Over four years, participants have stated in their final evaluations of the week that the Beaver Island experience changed how they thought about the role of literacy in science teaching. Collectively, the majority of comments suggest that participants were provided additional information by the PD that changed how they perceived their responsibilities and expectations as teachers. Additionally, the participants stated that integrating and combining themes (science and literacy) are easier than they had anticipated. They stated that they would be more selective when choosing texts, and include more than basic lab report writing in class. For instance anonymous responses state:

> The institute was a wonderful experience at a unique location. You provided relevant activities and usable information that could be taken back to the classroom and immediately implemented.
>
> I am able to more clearly explain how the new literacy standards are not as much about teaching ELA content in science class, but follow up on what is taught in ELA when reading and writing in science class.
>
> Now I speak with my colleagues often about how to integrate what they are doing with what we are working on … [this type of collaboration is not] limited to the colleagues who attended the institute with me.

We asked them, specifically, to describe what had changed in their thinking. Responses were positive and suggest that participants would engage in meaningful changes in their classroom practice. For instance:

> I am going to implement more writing and nonfiction reading into my classroom to engage my students. I was more traditional in the past with how I thought my students should learn concepts in science.
>
> This week was an eye opener in the sense that I feel I have what I need to bring new writing approaches into the lessons. QFT [Question Formulation Technique], Board meetings, story boards, nonfiction writing. Free choice notebook taking for gather data.
>
> I feel that before this experience literacy in the science classroom was strictly textbook or informational materials. This experience has opened my eyes to the idea of allowing more creativity to bridge the gap from literacy to science.
>
> It has not only changed my thinking about the role of literacy in science, but also every other subject. I realized that I don't write and reflect enough in my classroom. It was an invaluable lesson for me.
>
> Before this institute, literacy to me was something I had to cover with my students. After this experience, I realize that literacy and art are ways to enrich the learning experience for my students.
>
> My thinking has changed by showing me that it's okay to release control and allow students to explore their thinking. I now see that my gradebook can reflect this inquiry approach without needing as many worksheets.

Taken together, the comments from participants suggest that the institute is a transformative experience. Yet, understanding how teachers translate the experience into their own classroom gives us even more insights into these changes in their thinking, as shown in the next section.

Sample Units

Over the years we have done the Beaver Island Institute (2016–19), we have had a number of incredible colleagues who have been willing to make their practice public, sharing their final unit plans on our institute homepage (linked from this chapter's companion page). In particular, we highlight three units — and samples of student work — that have stood out over the years.

Clean Water: Creation and Access

>Chris Thoms and Jennifer Tapolcai
>Botwell Middle School
>Marquette, MI

Focusing on issues of local water quality and connecting it to conditions around the world, Chris Thoms and Jennifer Tapolcai invited students to read a novel, *A Long Walk to Water* by Linda Sue Park (2011), and then performed water quality sampling at various locations near the school and in the Marquette region, even visiting a local water treatment facility. Moreover, they explored water scarcity worldwide as well as implications of the Flint Water Crisis. These overlapping literacy and science goals aligned with the intent of our institute, and they were able to implement the unit with their students during the next school year. With all the data that they collected, as a final step in the unit, they asked students to design both physical and digital posters documenting their work on water scarcity.

We are fortunate enough to have an example of their students' work here, in the form of Stella Brunet's social media post, "Why Water?" (Figure 6.8). As they neared the end of the unit, Thoms and Tapolcai asked students to summarize and synthesize the main ideas from their inquiry into one social media post. As the adage goes, it is difficult to take big ideas and write small. Yet, in her post, Stella captured a key statistic and makes a clear call to action, set with white text, simple text on a grayscale image of waves hitting a shoreline. As is often the case in modern media, a URL is included at the end of the social media post, giving viewers the opportunity to find out more.

Figure 6.8 "Why Water" Project by Bothwell Middle School Student Stella Brunet (Image courtesy of Christopher Thoms and Stella Brunet)

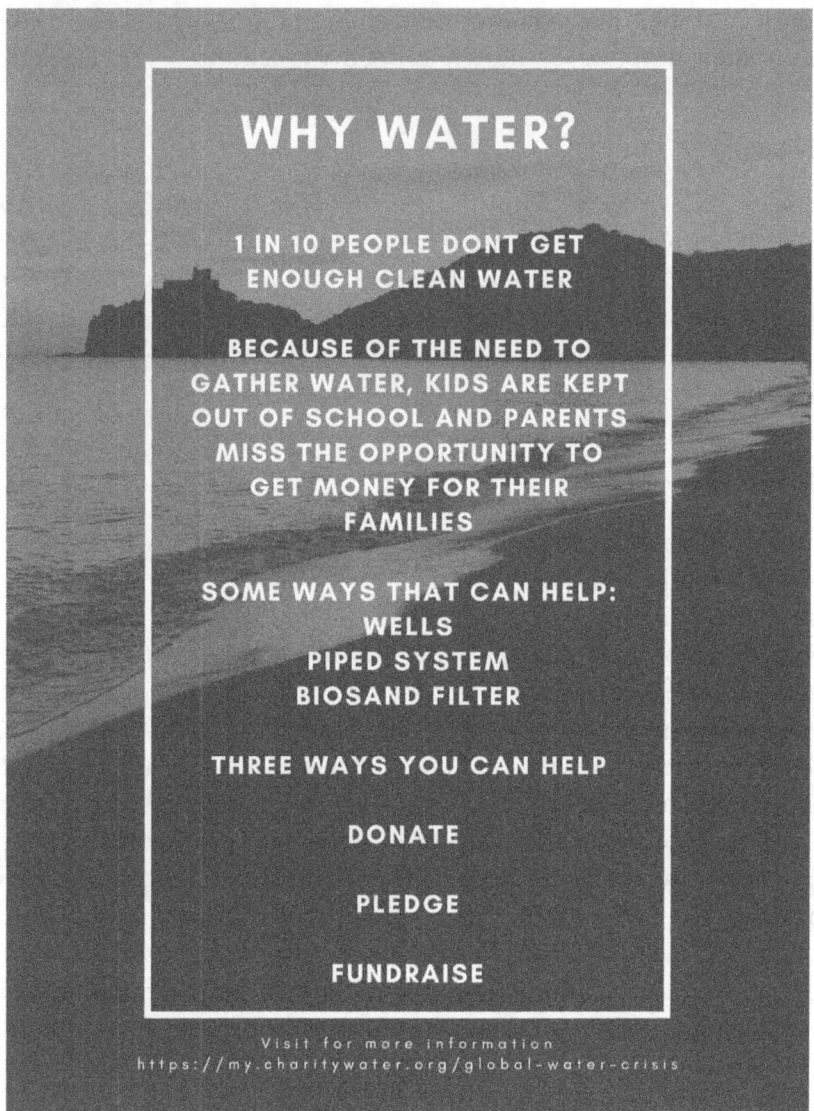

Interactions and Interdependence

Lindsay Thurman and Jamie Klausing
Highlander Way Middle School
Howell, MI

Considering the ways in which ecosystems are designed and the organisms within them are connected, Lindsay Thurman and Jamie Klausing invited students to research a particular organism native to their region in Michigan, starting with nature walks in and near their school and even exploring a game from Project WILD, "Oh Deer" (2014), that helped students to better understand habitat loss. Culminating in an exhibition where students presented their organisms, some using digital tools like PowerPoint and some on posterboard and tri-folds, the students were invited to walk from room to room in their wing of the school, visiting different ecosystems in each one. Moreover, they invited elementary students to visit, too, thus the middle school students had an authentic audience for their work. Again, Thurman and Klausing captured the main ideas of our institute, asking students to use disciplinary literacy practices in a strategic manner, communicating to an audience of other students.

Again, we are fortunate enough to have an example of their students' work here, in the form of Sophia Balla's poster about the Great Blue Heron (Figure 6.9). In her writing on the poster, she has taken a great deal of factual information about the heron and, for the most part, summarized it in her own words. Employing terms like "colonies," "overpopulated," "mutually beneficial," and "aquatic

Figure 6.9 "Great Blue Heron" Project by Highlander Way Middle School Student Sophia Balla (Image courtesy of Lindsay Thurman and Sophia Balla)

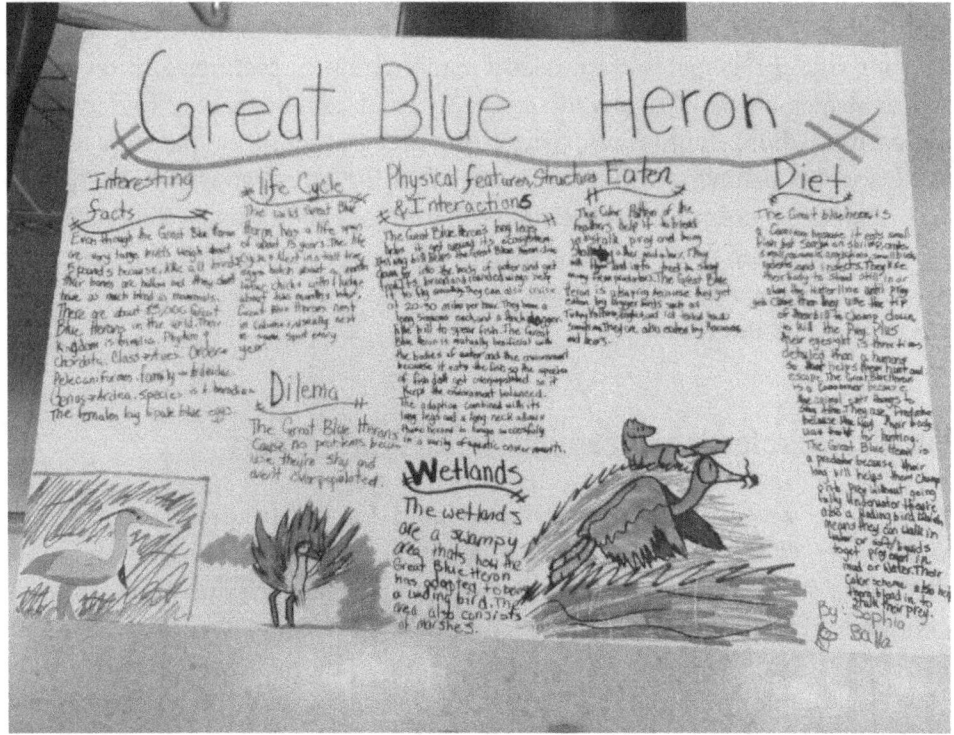

environment," she demonstrates her understanding of the animal. Moreover, she looked to other photos of the heron in order to hand draw her own interpretations of the bird in various states of rest and activity.

Species Integration and Management

Erin Wright
Eberwhite Elementary School
Ann Arbor, MI

Given that Eberwhite Elementary is located right next to a woodlot, Erin Wright's students have a natural laboratory right outside their schoolhouse steps. Throughout the year, students visit the woodlot, led by volunteers from a local nature conservancy group, and find out about the effects of invasive species, climate change, and active management. One of the projects they have pursued is to build and maintain habitats for house sparrows, since — as described by the Cornell Lab of Ornithology's All About Birds website — "[m]any people regard House Sparrows as undesirables in their yards, since they aren't native and can be a menace to native species" (Cornell Lab of Ornithology, n.d.). Capturing the spirit of the institute, Wright asked her students to generate their own questions about the species outside their window, and to document their learning in many ways.

And, one final time, we are fortunate enough to have an example of her students' work here, in the form of notes taken on the house sparrow (Figure 6.10). Wright's student, Isabel Sutton, uses a combination of pictures and text — as well as strategic uses of arrows to connect her ideas — as a way to document her learning about the house sparrow. She notes some of the peculiar aspects of the species' behavior and clearly notes that the house sparrow is "Aggressive, Attacks other birds and their nests." In this note-taking, we can see how Isabel is processing information about the house sparrow, making connections to what she knows about birds and their habitat, as well as other needs that species need for survival.

Rethinking Professional Learning for Scientific Literacy

As Troy has argued in a position statement for the International Literacy Association, "Democratizing Professional Growth with Teachers," there are some trends that can be noted in the ways that teacher education and professional development have changed in the past decade. Typical *professional development* "leaves educators feeling uninspired, de-professionalized, and at a loss for how to implement a number of disconnected strategies presented in a one-shot fashion" (Hicks & Sailors, 2018, p. 2). In contrast, what we have

Figure 6.10 "House Sparrow" Project by Eberwhite Elementary School Student Isabel Sutton (Image courtesy of Erin Wright and Isabel Sutton)

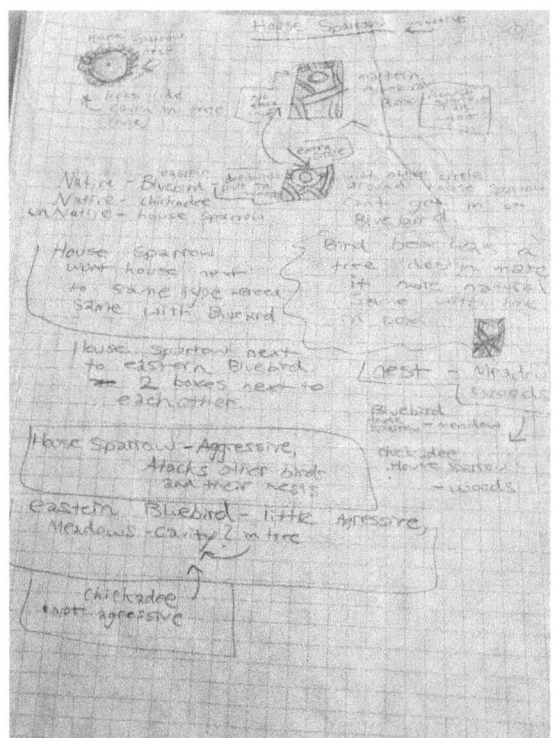

tried to do with the Beaver Island Institute is a model of professional learning that is "grounded in the notion that teachers are also learners, able to model and enact the processes of active inquiry, critical thinking, and problem solving with their students" (p. 3). We contend that immersive and sustained professional learning experiences can, potentially, lead teachers to substantive changes in their classroom practice.

As with any single professional learning experience, it is difficult to say with certainty that teachers who have participated in the Beaver Island experience have experienced significant change as a result. While the immediate results on the survey — as well as some promising practices observed in classrooms — all suggest that there are many positive outcomes, we know that the ultimate goal of any professional learning experience comes through in a greater sense of self-efficacy, deeper content knowledge, and new insights on teaching practices. Given the overall design, facilitation, and evaluation of the institute, we feel comfortable making the claim that these goals have been met.

In a broader sense, our goals in designing the institute in the ways we have are to, in turn, model for and encourage participants to engage in this kind of learning with their own students. Based on what we have learned from them, we are optimistic that, at least in some small way, this has happened in our participants' own classrooms, a theme that we discuss more in our conclusion.

Conclusion: Next Steps with the Next Gen Science Standards and Literacy Learning

We know that the NGSS, CCSS, and ISTE standards can feel overwhelming in and of themselves, when trying to integrate new ideas into existing curriculum, let alone moving to an entirely new mode of teaching through an inquiry-based approach. We've felt these pressures in our own teaching, as well as when facilitating events for other educators. And, despite these pressures — and the failures we may have encountered along the way — we continue to move our teaching toward inquiry.

As the title of our book articulates, the interwoven acts of asking questions, exploring ideas and data, and then writing about the entire process in various ways does, ultimately, lead students to become more effective, involved learners. Throughout the process of writing this book, the three of us have continued to share strategies with one another, pushing one another's thinking in different directions and encouraging each to try new ideas in our own classrooms.

At the start of the book, we outlined four ideas that have guided us:

1. Literacy and Science *Do* Mix;
2. Writing and Science are Both Process-Oriented;
3. Inquiry Matters;
4. Technology, Writing, and Science Are Intertwined.

As we conclude, we hope that these themes have become evident through the many examples that we have shared, both from Jeremy and Wiline's classrooms as well as our collaborative efforts at professional learning through the Beaver Island Institute. In all these spaces, we can expect to see students asking substantive questions, engaging in writing-to-learn, contributing to classroom conversations, interpreting and representing data in different ways and, ideally, using their creativity to share ideas in novel ways. As we wrap up, here are a few final words about each of these themes.

First, we are consistently mindful that teaching contexts vary. While we have described practices that work for us with middle school students and college undergraduates, as well as a population of teachers in Michigan, we know that the ideas in the book may need to be tweaked depending on the classroom, school, and community in which you work. In some ways, our strategies could fall prey to the trap of all books about teaching, "Yes, that worked for you, but ..." Our hope is that we have offered enough examples that can be adapted and extended. As we continue to refine our practices, we know that some ideas will hold up over time, and that some might change. We also are open to the experiences and insights of our colleagues, as we consider what we might do differently. That said, we know that everything presented in this book may not work in all contexts, and yet we are hopeful that the ideas can be a place to begin.

Second, and in a similar manner, as we return to the story that Jeremy shared in Chapter 1 about his math colleague, we know that there will still be some resistance. As our tweens and teens in our lives would remind us (with perhaps a slightly outdated slang expression), "haters gonna hate." Despite our best efforts to model effective teaching practices, to share student success, and offer support through collegial conversation, we know that not all of our colleagues will embrace these ideas. As a reader who has made it this far in the book, we know that you are likely to be someone who will try out at least a few of these ideas and, in turn, make your teaching practices public. With any luck, you will be able to share some ideas with at least one colleague who will, in turn, become more interested in collaborating with you as you develop these overlapping science and literacy practices.

Finally, we return to the core actions — and title — of the book: ask, explore, write. When we ask questions, and teach our students how to ask questions, amazing ideas can unfold. Second, as we welcome exploration and inquiry, we know that the path may sometimes not unfold as clearly as a traditional textbook or curricular planning document would suggest that it should. However, and at risk of sounding cliché, the journey is the destination, and inquiry can be an amazing excursion. And, with writing, we also run the risk of sharing platitudes, yet we see it as a trip, too. We can have students write in quick bursts, for sustained periods, and return to drafts of notebook entries as they transform their ideas into final drafts. Writing can be infused in the beginning, middle, or end of a class period, a unit of study, a lab experience, or any other teaching moment. Our hope, again, is that we have offered some insights on how we do this work.

Potential Next Moves for Teachers

As the adage goes, the person in the classroom doing the most work is the one doing the most learning. We no longer want our schools, and especially our science classrooms, to be the place where students go to watch teachers work. Without actively pursuing their own inquiry, the chances are pretty good that students will remain in a passive role, memorizing facts, vocabulary, and formulas just long enough to pass the test at the end of a unit. So, if you are looking for just one task, one idea to get started, here is what we recommend:

- For the teacher just wanting to get students writing in science, go back to Chapter 3 and look at Sidebars 3.1, "Visual Thinking Strategy Useful for Analyzing Infographics" and 3.3, "Questions to Prompt Student Thinking Around Infographics." These will help to get students interpreting infographics and, in turn, begin to write about them in substantive ways.
- For the teacher who is planning their next unit, consider the ways that science notebooks (Chapter 2) or whiteboarding (Chapter 4) can fit into the unit you are teaching. While it may take some time to plan when and where these activities will fit in, we encourage you to try them out at least two or three times throughout the unit and see what a difference they can make.
- For the teacher who wants to get students involved in a project, go back to Chapter 6 and look at Activity 1, "Snakeboards (Observational Design) or 5, "Invasive Species Removal." Both of these activities, or ones of similar design, are fairly easy to set up in an area near your school, and have the added benefit of being highly engaging.
- For the teacher who is looking to bring in literature into their classrooms, review our brief recommendations in Chapter 5, or visit the NSTA great books website (link available on companion page).

No matter what the next moves are that you make, we reiterate the main themes — and key words in our title — here, one final time. By inviting students to ask their own questions (and scaffolding them in the process of doing so), we can support their active inquiry. By exploring scientific phenomena, they can begin to answer their own questions, based on the data that they have gathered. Finally, when we ask them to write — whether short spurts, longer takes, or through multimodal forms such as infographics — we provide students with new opportunities for expressing their understanding of scientific concepts. From what we have experienced in our classrooms, and what we see happening with like-minded educators in their classrooms, we believe that the process of asking, exploring, and writing can help our students adopt the skills and dispositions that they need to be scientifically literate, now and in the future.

References

Achieve Inc. (2013a). The Next Generation Science Standards. Retrieved March 25, 2013, from www.nextgenscience.org/next-generation-science-standards

Achieve Inc. (2013b, May). Appendix M – Connections to the Common Core State Standards for Literacy in Science and Technical Subjects. Retrieved from www.nextgenscience.org/sites/default/files/Appendix%20M%20Connections%20to%20the%20CCSS%20for%20Literacy_061213.pdf

Achieve Inc., Next Generation Science Standards, Next Gen Storylines, & STEM Teaching Tools. (2016, September). *Using Phenomena in NGSS-designed Lessons and Units*. Retrieved from www.nextgenscience.org/sites/default/files/Using%20Phenomena%20in%20NGSS.pdf

Achieve, Inc. (2018). About Achieve. Retrieved August 24, 2019, from *Achieve website*: www.achieve.org/about-us

Achieve, Inc. (n.d.). NGSS Fact Sheet. Retrieved August 24, 2019, from www.nextgenscience.org/resources/ngss-fact-sheet

American Modeling Teachers Association. (n.d.). Synopsis of Modeling Instruction. Retrieved June 19, 2019, from https://modelinginstruction.org/sample-page/synopsis-of-modeling-instruction/

Baker, F. (2017, May 30). Media Literacy: How to Close Read Infographics. Retrieved January 28, 2019, from *MiddleWeb website*: www.middleweb.com/34963/media-literacy-how-to-close-read-infographics/

Bazerman, C., Little, J., & Bethel, L. (2005). *Reference Guide to Writing across the Curriculum*. West Lafayette, Ind: Parlor Press.

Bean, J. C. (2011). *Engaging Ideas: The Professor's Guide to Integrating Writing, Critical Thinking, and Active Learning in the Classroom* (2 ed.). San Francisco: Jossey-Bass.

Blaettler, K. G. (2018, May 17). The Difference between Charts & Graphs. Retrieved January 28, 2019, from *Sciencing website*: https://sciencing.com/difference-between-charts-graphs-7385398.html

Bolduc, N. (2019, July 8). 5 Tips for Blending the Question Formulation Technique (QFT) with NGSS. Retrieved July 25, 2019, from *Teaching Channel website*: www.teachingchannel.org/tch/blog/5-tips-blending-question-formulation-technique-qft-ngss

Bryson, B. (2006). *A Walk in the Woods: Rediscovering America on the Appalachian Trail* (2nd ed.). Bismarck, ND: Anchor.

Carter, M., Ferzli, M., & Wiebe, E. N. (2007). Writing to Learn by Learning to Write in the Disciplines. *Journal of Business and Technical Communication*, 21(3), 278–302. doi:https://doi.org/10.1177/1050651907300466

Chalabi, M. (2019, September 18). High School Students Who Smoke Cigarettes or e-cigarettes. Retrieved September 20, 2019 from *The Guardian*.

Cornell Lab of Ornithology. (n.d.). House Sparrow Overview. Retrieved August 30, 2019, from *All About Birds, website*: www.allaboutbirds.org/guide/House_Sparrow/overview

Council of Chief State School Officers, & National Governors Association Center for Best Practices. (2010). English Language Arts Standards. Retrieved December 20, 2015, from www.corestandards.org/ELA-Literacy/L/language-progressive-skills/

Daniels, H. (2004). *Subjects Matter: Every Teacher's Guide to Content-area Reading*. Portsmouth, NH: Heinemann.

Daniels, H., Zemelman, S., & Steineke, N. (2007). *Content-area Writing: Every Teacher's Guide*. Portsmouth, NH: Heinemann.

deLacy, M. (2017, November 16). Where Do NGSS and ELA Standards Intersect? Retrieved July 23, 2018, from www.knowatom.com/blog/where-do-ngss-and-ela-standards-intersect

Derry, G. N. (2002). *What Science is and How It Works* (Edition Unstated edition). Princeton, N.J. Oxford: Princeton University Press.

Elliott, L. A., Jaxon, K., & Salter, I. (2016). *Composing Science: A Facilitator's Guide to Writing in the Science Classroom*. New York, NY: Berkeley, CA: Teachers College Press.

Fries-Gaither, J. (2017). *Notable Notebooks: Scientists and Their Writings* (Ntb ed.). Arlington, Virginia: NSTA Kids.

Gardner, T. (2011). Designing Writing Assignments. National Council of Teachers of English. Urbana, IL. Retrieved from https://wac.colostate.edu/books/gardner/

George, J. C. (2004). *My Side of the Mountain* (First Printing edition). Princeton, N.J.: Puffin Books.

Gigante, M. (2016). Navigating the Next Generation Science Standards: Implications and Implementation for Faculty in Writing and the Sciences. *Teaching/Writing: The Journal of Writing Teacher Education*, 5, (1). Retrieved from http://scholarworks.wmich.edu/wte/vol5/iss1/6

Grant, M. C., Fisher, D., & Lapp, D. (2015). *Reading and Writing in Science: Tools to Develop Disciplinary Literacy* (Second edition). Thousand Oaks, CA: Corwin.

Gray, D. (2011, May 17). 6-8-5. Retrieved August 30, 2019, from *Gamestorming website*: https://gamestorming.com/6-8-5s/

Gruwell, E., & The Freedom Writers. (1999). *The Freedom Writers Diary: How a Teacher and 150 Teens Used Writing to Change Themselves and the World around Them*. New York, NY: Broadway Books.

Gutkind, L. (2012). *You Can't Make this Stuff Up: The Complete Guide to Writing Creative Nonfiction–From Memoir to Literary Journalism and Everything in Between*. Boston, MA: Hachette Books.

Heick, T. (2013, May 8). 10 Brilliant Examples of Sketch Notes: Notetaking for the 21st Century. Retrieved August 24, 2019, from *TeachThought website*: www.teachthought.com/literacy/10-brilliant-examples-of-sketch-notes-notaking-for-the-21st-century/

Hermann, R. S. (2007). Evolution as a Controversial Issue: A Review of Instructional Approaches. *Science & Education*, 17(8), 1011–1032. doi: https://doi.org/10.1007/s11191-007-9074-x

Hiaasen, C. (2005). *Hoot*. New York, NY: Yearling.

Hicks, T. (2013). *Crafting Digital Writing: Composing Texts across Media and Genres*. Portsmouth, NH: Heinemann.

Hicks, T., Bruner, J., & Kaya, T. (2017). Implementation of Blogging as an Alternative to the Lab Report. *International Journal of Engineering Education*, 33(4), 1257–1270.

Hicks, T., & Sailors, M. (2018). *Democratizing Professional Growth with Teachers: From Development to Learning* (No. 9437). Retrieved from *International Literacy Association website*: https://literacyworldwide.org/docs/default-source/where-we-stand/ila-democratizing-professional-growth-with-teachers.pdf

Hicks, T., & Steffel, S. (2012). Learning with Text in English/Language Arts. In T. L. Jetton & C. Shanahan (Eds.), *Adolescent Literacy in the Academic Disciplines General Principles and Practical Strategies* (pp. 120–153). Retrieved from www.guilford.com/cgi-bin/cartscript.cgi?page=pr/jetton2.htm&dir=edu/lit&cart_id=854361.19979

Hoffer, W. W. (2009). *Science as Thinking: The Constants and Variables of Inquiry Teaching, Grades 5-10*. Portsmouth, NH: Heinemann.

Hood, D. (2010, February 17). Writing Creative Nonfiction. Retrieved August 13, 2018, from *Find Your Creative Muse website*: https://davehood59.wordpress.com/2010/02/17/writing-creative-nonfiction/

Housen, A., & Yenawine, P. (n.d.). What is VTS? Retrieved from www.castellaniartmuseum.org/assets/Images/Documents-pdfs-applications/All-Lessons-VTS-Resourse.pdf

Hyler, J., & Hicks, T. (2014). *Create, Compose, Connect! Reading, Writing, and Learning with Digital Tools*. New York, NY: Routledge.

Hyler, J., & Hicks, T. (2017). *From Texting to Teaching: Grammar Instruction in a Digital Age*. New York, NY: Routledge.

International Society for Technology in Education. (2016). ISTE Standards for Students. Retrieved December 7, 2018, from www.iste.org/standards/for-students

Jackson, J., Dukerich, L., & Hestenes, D. (2008). Modeling Instruction: An Effective Model for Science Education. *Science Educator*, 17(1), 10–17.

Jensen, M. S., & Finley, F. N. (1995). Teaching Evolution Using Historical Arguments in a Conceptual Change Strategy. *Science Education*, 79(2), 147–166. doi:https://doi.org/10.1002/sce.3730790203

Jensen, M. S., & Finley, F. N. (1996). Changes in Students' Understanding of Evolution Resulting from Different Curricular and Instructional Strategies. *Journal of Research in Science Teaching*, 33(8), 879–900. https://doi.org/10.1002/(SICI)1098-2736(199610)33:83.0.CO;2-T

Jensen, M. S., & Finley, F. N. (1997). Teaching Evolution Using a Historically Rich Curriculum & Paired Problem Solving Instructional Strategy. *The American Biology Teacher*, 59(4), 208–212. doi: https://doi.org/10.2307/4450287

Jetton, T. L., & Shanahan, C. (Eds.). (2012). *Adolescent Literacy in the Academic Disciplines: General Principles and Practical Strategies* (1st ed.). New York, NY: The Guilford Press.

Juarez, K., Thompson-Grove, G., & Feicke, K. (2007). The Charette Protocol. Retrieved from http://schoolreforminitiative.org/doc/charrette.pdf

Klein, P. D., & Boscolo, P. (2016). Trends in research on writing as a learning activity. *Journal of Writing Research*, 7(3), 311–350. doi: https://doi.org/10.17239/jowr-2016.07.03.01

Kolb, L. (2017). *Learning First, Technology Second: The Educator's Guide to Designing Authentic Lessons*. Portland, Oregon: International Society for Technology in Education.

Lane, B. (2003). *51 Wacky We-search Reports: Face the Facts with Fun*. Shoreham, VT: Discover Writing Press.

Lent, R. C. (2015). *This is Disciplinary Literacy: Reading, Writing, Thinking, and Doing… Content Area by Content Area* (1 edition). Thousand Oaks, CA: Corwin.

Lile, S. (n.d.). 44 Types of Graphs Perfect for Every Top Industry. Retrieved August 31, 2019, from *Visual Learning Center by Visme website*: https://visme.co/blog/types-of-graphs/

Linton, D. L., Pangle, W. M., Wyatt, K. H., Powell, K. N., Sherwood, R. E., & Momsen, J. (2014). Identifying Key Features of Effective Active Learning: The Effects of Writing and Peer Discussion. *CBE—Life Sciences Education*, 13(3), 469–477. doi: https://doi.org/10.1187/cbe.13-12-0242

Martix, S., & Hodson, J. (2014). Teaching with Infographics: Practising New Digital Competencies and Visual Literacies. *Journal of Pedagogic Development*, Retrieved from http://uobrep.openrepository.com/uobrep/handle/10547/335892

McComas, W. F., & Nouri, N. (2016). The Nature of Science and the Next Generation Science Standards: Analysis and Critique. *Journal of Science Teacher Education*, 27(5), 555–576. doi:https://doi.org/10.1007/s10972-016-9474-3

McKagan, S., & McPadden, D. (2017, October 26). Best Practices for Whiteboarding in the Physics Classroom. Retrieved June 9, 2019, from *PhysPort website*: www.physport.org/recommendations/Entry.cfm?ID=101319

Merriam-Webster Online Dictionary. (n.d.a). Definition of GRAPH. Retrieved January 28, 2019, from www.merriam-webster.com/dictionary/graph

Merriam-Webster Online Dictionary. (n.d.-b). Definition of TABLE. Retrieved January 28, 2019, from www.merriam-webster.com/dictionary/table

National Forum on Education Statistics. (2010). *Forum Guide to Data Ethics (NFES 2010-801)*. Washington, DC: U.S. Department of Education, National Center for Education Statistics.

National Writing Project. (n.d.). About NWP – National Writing Project. Retrieved February 18, 2012, from www.nwp.org/cs/public/print/doc/about.csp

Nguyen, N. (2016, May 31). 23 Bullet Journal Ideas that Are Borderline Genius. Retrieved August 24, 2019, from *Nifty on BuzzFeed website*: www.buzzfeed.com/nicolenguyen/genius-ways-you-can-customize-your-bullet-journal

Pareyt, B., Talhaoui, F., Kerckhofs, G., Brijs, K., Goesaert, H., Wevers, M., & Delcour, J. A. (2009). The Role of Sugar and Fat in Sugar-snap Cookies: Structural and Textural Properties. *Journal of Food Engineering*, 90(3), 400–408. doi: https://doi.org/10.1016/j.jfoodeng.2008.07.010

Park, L. S. (2011). *A Long Walk to Water* (Reprint edition). Boston: HMH Books for Young Readers.

Paulsen, G. (2009). *Hatchet*. New York, NY: Simon and Schuster.

Pennypacker, S. (2019). *Pax*. New York, NY: Balzer and Bray.

Posthuma, E. (2014, January 26). Whiteboarding Strategies [Text]. Retrieved June 9, 2019, from *Chemical Education Xchange website*: www.chemedx.org/article/whiteboarding-strategies

Project WILD, & Cornell Institute for Biology Teachers. (2014). Oh Deer. Retrieved September 22, 2019, from https://blogs.cornell.edu/cibt/labs-activities/labs/oh-deer-mary-bowman/

Richtel, M., & Kaplan, S. (2019, September 19). Vaping Illnesses Increase to 530 Probable Cases, c.d.c. Says. *The New York Times*. Retrieved from www.nytimes.com/2019/09/19/health/vaping-cdc.html

Right Question Institute. (2019). What is the QFT? Retrieved August 30, 2019, from *Right Question Institute website*: https://rightquestion.org/what-is-the-qft/

Sapolsky, R. M. (2017). *Behave: The biology of humans at our best and worst*. New York, New York, NY: Penguin Press.

Shanahan, T., & Shanahan, C. (2008). Teaching Disciplinary Literacy to Adolescents: Rethinking Content-area Literacy. *Harvard Educational Review*, 78(1), 40–59.

Sunrise Science. (2018, May 11). *Free Websites for Finding NGSS Anchoring Phenomena*. Retrieved June 9, 2019, from *Sunrise Science Blog website*: http://sunrisescience.blog/free-websites-ngss-anchoring-phenomena/.

Trouilloud, W., Delisle, A., & Kramer, D. L. (2004). Head Raising during Foraging and Pausing during Intermittent Locomotion as Components of Antipredator Vigilance in Chipmunks. *Animal Behaviour*, 67(4), 789–797. doi: https://doi.org/10.1016/j.anbehav.2003.04.013

Tufte, E. R. (2001). *The Visual Display of Quantitative Information* (2nd ed.). Cheshire, Conn: Graphics Press.

Turner, K. H., & Hicks, T. (2016). *Argument in the Real World: Teaching Adolescents to Read and Write Digital Texts*. Portsmouth, NH: Heinemann.

United States Department of Agriculture. (2013). USDA Farm to School Census Infographic [Photo]. Retrieved from www.flickr.com/photos/usdagov/10410338223/

van Dijk, A. M., Gijlers, H., & Weinberger, A. (2014). Scripted Collaborative Drawing in Elementary Science Education. *Instructional Science*, 42(3), 353–372. doi: https://doi.org/10.1007/s11251-013-9286-1

For Product Safety Concerns and Information please contact our EU representative GPSR@taylorandfrancis.com
Taylor & Francis Verlag GmbH, Kaufingerstraße 24, 80331 München, Germany

www.ingramcontent.com/pod-product-compliance
Lightning Source LLC
Chambersburg PA
CBHW080937300426
44115CB00017B/2860